Practical Acoustic Emission Testing

The Japanese Society for Non-Destructive Inspection

Practical Acoustic Emission Testing

Springer

The Japanese Society for Non-Destructive Inspection
Tokyo, Japan

Original Japanese edition published by The Japanese Society for Non-Destructive Inspection.
Acoustic Emission Testing I
Copyright © The Japanese Society for Non-Destructive Inspection 2006

ISBN 978-4-431-56641-0   ISBN 978-4-431-55072-3   (eBook)
DOI 10.1007/978-4-431-55072-3

Printed on acid-free paper

This Springer imprint is published by SpringerNature
The registered company is Springer Japan KK.

Preface

Acoustic emission (AE) measurement is a technique for detecting an elastic wave that is generated by the occurrence of microscale defects. Phenomena resulting from microscale defects can thus be readily detected by AE. Recently, the accidental failure of various types of structures, resulting in disasters and injury to people in many cases, has been reported as posing a threat to human safety. In most cases, the durability and the load-bearing capacity of facilities or structures are carefully ensured against microscale defects (structural flaw and cracks). However, in some recent accidents, it has been found that microscale defects can lead to dangerous conditions of structures.

Therefore, the establishment of "structural diagnosis" is in urgent demand. In this respect, AE techniques are known to be promising for detecting microscale defects and the analysis of "fracture phenomena." Traditionally, the usefulness of such techniques for predicting rockfalls has been known in mines worldwide. The technique received great attention when it was applied to a pressure test of a rocket motor case (rocket body) of a Polaris missile in the United States. Later, AE testing was standardized as a nondestructive inspection technique for detecting defects in pressure vessels and tanks. This became the motivation for current AE testing.

In recent years, applications of AE measurements have been extended from the fields of metal and mechanical engineering to those of civil and chemical engineering, resulting in the establishment of practical inspection in many fields. Thus the education of non-destructive testing (NDT) technicians in AE has become an important issue.

This book was originally prepared for NDT technicians who need to learn practical acoustic emission testing based on level 1 of ISO 9712 (Non-destructive testing –Qualification and Certification of personnel) by the research and technical committee on AE of the JSNDI (The Japanese Society for Non-Destructive Inspection). The book was also selected as an essential reference of ISO/DIS 18436-6 (Condition monitoring and diagnostics of machines –Requirements for training and certification of personnel, Part 6: Acoustic Emission).

The editors (see below) and authors of the book are key members of the research and technical committee on AE of the JSNDI. This 30-year-old committee has consecutively held the International AE symposium (IAES) every 2 years.

Against this background – the roles of NDT technicians – the principles of AE measurement are explained clearly followed by signal processing, algorithms for source location, measurement devices, applicability of testing methods, and measurement cases. The authors hope that this book will play a key role in AE education and study in all fields of engineering.

Editorial Board
Chair
Yoshihiro Mizutani, Tokyo Institute of Technology
Members
Manabu Enoki, The University of Tokyo
Hidehiro Inaba, Fuji Ceramics Corporation
Hideyuki Nakamura, IHI Inspection and Instrumentation Co., Ltd.
The late Masaaki Nakano, Chiyoda Corporation
Mitsuhiro Shigeishi, Kumamoto University
Tomoki Shiotani, Kyoto University
Shin-ichi Takeda, Japan Aerospace Exploration Agency
Shigenori Yuyama, Nippon Physical Acoustics, Ltd.

Opening Figures

1. Applications

Remote monitoring of damages in a suspension bridge (www.mistrasgroup.com with permission)

Integrity evaluation of railway concrete bridge piers

Integrity evaluation of a pressure vessel

Rock failure monitoring with a remote system

Evaluation of fracture characteristics of a concrete specimen

Damage monitoring of an aircraft during a structural test (www.mistrasgroup.com with permission)

Damage monitoring of a rocket motor case (www.mistrasgroup.com with permission)

Evaluation of corrosion damage during tank bottom testing

Detection and evaluation of a valve leak (www.mistrasgroup.com with permission)

2. Monitoring Systems and Software

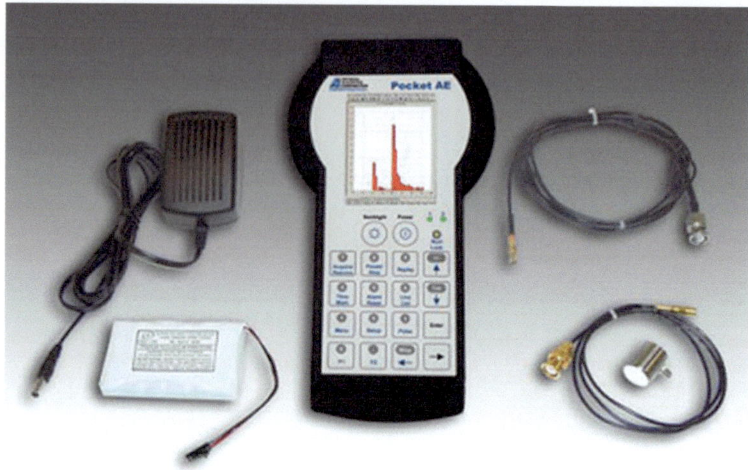

Portable acoustic emission system

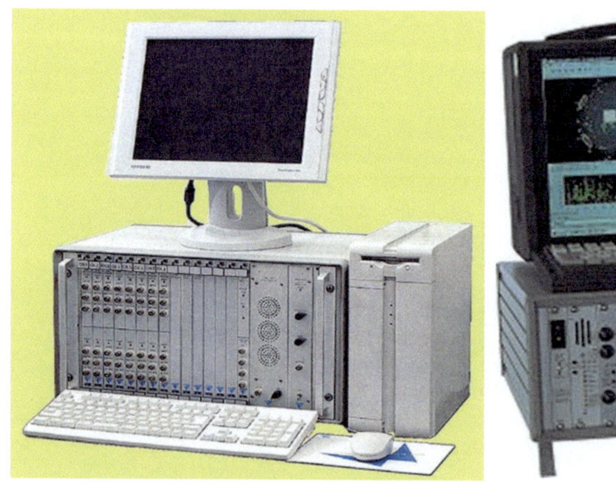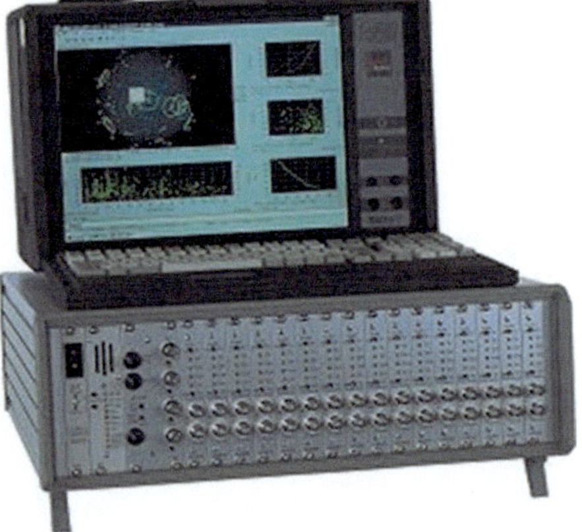

Multi-channel acoustic emission systems

Contents

List of Contributors

Manabu Enoki The University of Tokyo, Tokyo, Japan

Hidehiro Inaba Fuji Ceramics Corporation, Fujinomiya, Japan

Yoshihiro Mizutani Tokyo Institute of Technology, Tokyo, Japan

Hideyuki Nakamura IHI Inspection & Instrumentation Co., Ltd.,, Yokohama, Japan

Masaaki Nakano (deceased) Chiyoda Corporation, Yokohama, Japan

Masayasu Ohtsu Kumamoto University, Kumamoto, Japan

Mitsuhiro Shigeishi Kumamoto University, Kumamoto, Japan

Tomoki Shiotani Kyoto University, Kyoto, Japan

Sunao Sugimoto Japan Aerospace Exploration Agency, Tokyo, Japan

Shin-ichi Takeda Japan Aerospace Exploration Agency, Tokyo, Japan

Shigenori Yuyama Nippon Physical Acoustics, Ltd., Tokyo, Japan

Chapter 1
Roles and Safety/Health of Technicians Involved in Non-destructive Testing

Hideyuki Nakamura

Abstract In this chapter, roles of level 1 technicians involved in non-destructive testing (NDT technicians) as specified by ISO9712 are presented. Safety and health for NDT technicians are also demonstrated in this chapter.

Keywords Roles of level 1 technicians • Safety and health

1.1 Importance of Non-destructive Testing

Non-destructive testing (NDT) refers to tests conducted to non-destructively evaluate the soundness and internal condition of materials, equipment, and structures. In addition to AE testing (AT), non-destructive tests include visual testing (VT), radiographic testing (RT), ultrasonic testing (UT), magnetic testing (MT), penetration testing (PT), eddy-current testing (ET), strain testing (ST), leak testing (LT), and infrared thermography testing (TT). An optimal test is selected from among these tests according to test objects and purposes. Several tests are combined to use when high quality levels are required. The purposes of NDT include establishing manufacturing techniques and reducing manufacturing costs, but the most important purpose is to prevent any serious accident that causes injury to people or causes environmental contamination when a product breaks during use. To prevent the breakage of a product, it is important to verify the validity of a product design and thoroughly ensure the product strength during production. It is also important to detect degradation such as cracking and wall thinning of the product that exceeds an acceptable criterion for the design and to repair the product as needed to ensure quality during use. NDT is a method of detecting cracking and wall thinning that reduce strength of equipment during production and use of the product, as well as a method of evaluating the product's integrity; NDT is thus a critical means for ensuring the reliability of the product. An NDT technician's lack of expertise will result in improper testing and increase the risk of a serious disaster resulting from oversight or underestimation in the case of cracking or thinning. Therefore, NDT

H. Nakamura (✉)
IHI Inspection & Instrumentation Co., Ltd., Yokohama, Japan
e-mail: hideyuki_nakamura@iic.ihi.co.jp

© Springer Japan 2016
The Japanese Society for Non-Destructive Inspection, *Practical Acoustic Emission Testing*, DOI 10.1007/978-4-431-55072-3_1

must be conducted by a sufficiently skilled technician. Past investigations of serious accidents show cases in which false reporting and falsification of test results contributed to such accidents. The NDT technician must recognize the importance of his/her duties, be aware of his/her social responsibility, and carry out the duty with a sense of ethics and pride.

1.2 Role of a Level 1 Technician as Specified by ISO9712

The qualification and certification systems for NDT technicians have been established on the basis of the ISO international standard ISO9712. According to this certification system, if an technician passes a qualification test and satisfies the given conditions, he/she will be certified as a Level 1, 2, or 3 technician. An NDT technician's role depends on the NDT level, and duties that can be carried out at each level and the required ability are defined. The duties of NDT level 1 technician are given as follows.

1.2.1 Duties Acknowledged for NDT Level 1 Technicians

Any technician certified as an NDT Level 1 technician must be able to carry out the following duties under the supervision of a Level 2 or Level 3 technician.

(a) Prepare NDT equipment for AE testing
    The technician can install and adjust AE devices, sensors, preamplifiers, and cables.
(b) Operate NDT equipment
    For AE testing, the technician can conduct tasks such as device setup.
(c) Implement NDT for AE testing
    The technician can acquire and analyze data under the measurement conditions of the procedure.
(d) Record NDT results
    In AE testing, the technician can record data and results of data analysis.
(e) Classify and report NDT results in accordance with documented acceptance criteria
    In AE testing, the technician can determine acceptance/rejection, classify grades, and report the results in accordance with documented acceptance criteria.

1.2.2 Responsibilities of Level 1 Technicians

The NDT Level 1 technician is not accountable for the selection of NDT methods or techniques. The Level 1 technician assumes the duties mentioned above, but the preparation of NDT instructions, equipment adjustment and calibration, and interpretation and evaluation of NDT results that the Level 1 technician does not have to conduct are the duties of Level 2 or higher Level technicians. The selection of NDT methods or techniques is the responsibility of a Level 3 technician. It is important for all technicians to fully understand their own duties in detail and to note that their duties do not differ from a set of specified duties.

1.3 Health and Safety for NDT Technicians

In recent years, AE testing has been increasingly carried out in petroleum refining facilities and energy plants. Because there are many potential dangers in such work environments, careful attention must be paid to safety at these kinds of sites.

1.3.1 Caution in Equipment Handling

When technicians handle measurement devices, they must carefully read instruction manuals for the devices and avoid the occurrence of electric shock and fires caused by faulty handling. Particular attention should be paid to the following general matters.

(a) Do not connect/disconnect power and signal plugs with the power on

In AE testing, if a power plug or signal cable is mated or demated when an AE measurement device is powered on, there are risks of device failure, electric shock, and spark-induced fire. It is critical to never make this mistake.

(b) Do not disassemble an AE measurement device

The AE device and its peripheral equipment include built-in power sources and exposed electric wires. Therefore, removing their covers or disassembling the devices may cause electric shock or fire. Never disassemble the devices.

(c) Always connect the AE measurement device to an earthed wire

The insulation of an AE device and its peripheral equipment may reduce upon the ingress of dust or moisture into the devices, thereby causing a risk of electric leakage and shock. Therefore, it is necessary to connect a device to an earth wire before using it.

1.3.2 Work at High Elevations

In the AE testing of a large structure, technicians sometimes ascend to an elevated spot (height exceeding 2 m) to install a sensor or other device. In this case, it is desirable to place a scaffold in advance. When a stepladder or ladder is used, it is necessary to firmly fix the ladder to avoid the risk of falling. A technician must not ascend/descend to the spot while carrying equipments by hand. Further, when the technician works at an elevated spot, they must wear a safety belt and ensure that they are held by a rope in case of a fall, while paying careful attention to their activities.

1.3.3 Prevention of Explosions

Plants such as oil refining and gas production facilities, where no flame is permitted, may be required to use explosion-proof equipments. Before the implementation of AE testing at these sites, it is necessary to discuss the specifications of measurement devices with the people in charge to prevent an explosion. In actual activities, laws/regulations and safety provisions in each plant shall have priority over general precautions for the above-mentioned measurement.

Before carrying out AE testing, it is also important to provide an opportunity for a preliminary review of the test, determine danger factors, and establish counter-measures. It is important to obtain to a wide range of opinions and reassess the work environment from various perspectives.

Chapter 2
Principles of the Acoustic Emission (AE) Method and Signal Processing

Masayasu Ohtsu, Manabu Enoki, Yoshihiro Mizutani, and Mitsuhiro Shigeishi

Abstract Physical principles of the Acoustic Emission (AE) and the signal processing are presented in this chapter. The mechanism inducing AE waves are explained in comparison with that of an interpolate earthquake. Types and characteristics of AE sources are also explained. Fundamentals of AE propagation in solids are discussed. As for the evaluation several promising AE parameters and such AE source location techniques as 1D, 2D and zonal location are explained.

Keywords AE sources • Wave propagation • AE parameters • Source location

2.1 Principles of the AE Method

Masayasu Ohtsu and Yoshihiro Mizutani

When an external force is applied to a solid material, the material deforms. In the case of low stress due to a small external force, the deformed material elastically recovers to its original shape upon unloading (Fig. 2.1a). Such deformation is called elastic deformation. In the elastic range, the external force leads to accumulation of energy inside the material as strain energy.

There is a limit for energy accumulation in the solid material. If the strain energy stored in the material due to the external force reaches the limit, it is released and results in plastic deformation. That is, even when the external force is unloaded, the material cannot recover its original shape (Fig. 2.1b).

M. Ohtsu (✉) • M. Shigeishi
Department of Civil Engineering, Kumamoto University, Kumamoto, Japan
e-mail: ohtsu@gpo.kumamoto-u.ac.jp; shigeishi@civil.kumamoto-u.ac.jp

M. Enoki
Department of Materials Engineering, The University of Tokyo, Tokyo, Japan
e-mail: enoki@rme.mm.t.u-tokyo.ac.jp

Y. Mizutani
Department of Mechanical Sciences and Engineering, Tokyo Institute of Technology, Tokyo, Japan
e-mail: ymizutan@mes.titech.ac.jp

© Springer Japan 2016
The Japanese Society for Non-Destructive Inspection, *Practical Acoustic Emission Testing*, DOI 10.1007/978-4-431-55072-3_2

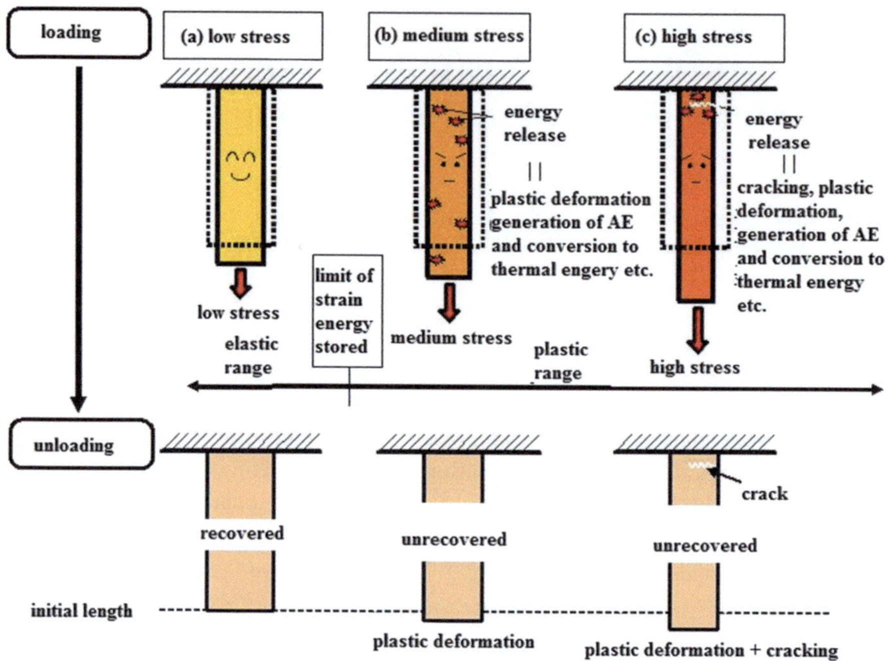

Fig. 2.1 Generation of AE due to strain energy release

If even larger external force is applied, cracking usually occurs in the material. In this case, the strain energy accumulated inside the material is consumed by the generation and growth of cracks (Fig. 2.1c).

When the solid material is deformed or cracked, it consumes strain energy. Thus, most of the strain energy is released. Simultaneously, remaining energy is consumed to generate sound and heat. A phenomenon in which sound is generated by the energy release is referred to as acoustic emission (AE), and it is sometimes described as the "scream" of the material under stress.

Sound is a phenomenon of energy release in air. In principle, because of cracking inside the material, elastic waves are generated and propagate through the material. The phenomenon is analogous to an earthquake. In other words, AE can be referred to as a "micro-earthquake" in a solid material.

In summary, AE is defined as a phenomenon in which strain energy accumulated in a solid is released because of deformation or cracking, and thus generates elastic waves. AE waves are detected usually at the surface of the material by a sensor as illustrated in Fig. 2.2.

Elastic waves generated by cracking propagate through the material and are detected by an AE sensor (vibration-to-electrical signal conversion element) placed on the surface. In this case, some portions of surface vibrations are released to the air as sonic waves and can possibly be heard as a breaking sound.

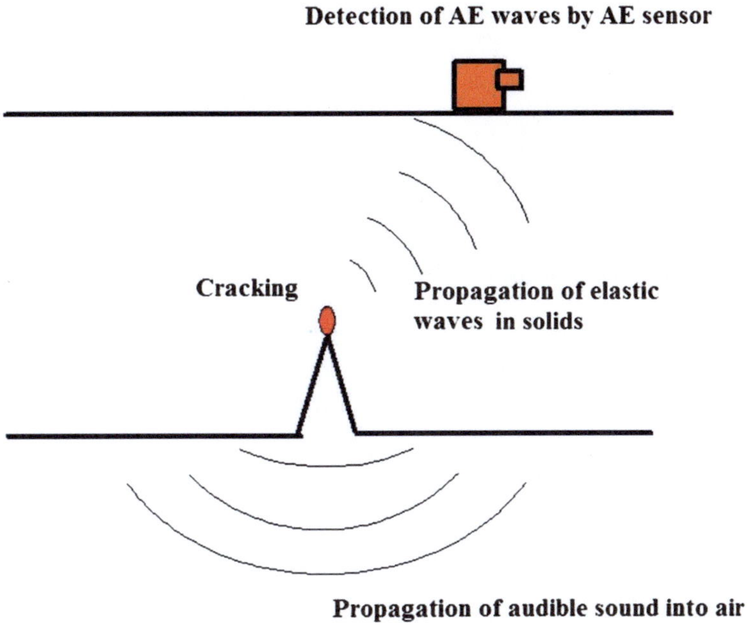

Fig. 2.2 Generation and detection of AE waves

To explain the source mechanisms of AE phenomena as an analogy of an earthquake, the focal mechanisms of an earthquake are illustrated in Fig. 2.3. A fault slip forms on a plate boundary between an oceanic plate and continental plate of the Earth's crust. As a result, a sliding failure or shear failure is nucleated. In the case of AE phenomena, two types of dislocations in Fig. 2.4 can be referred to as crack motions at a crack surface. A sliding or slip crack-motion (shear crack) corresponds to "in-plane shear dislocation", and tensile dislocation (tensile crack) is referred to as "opening dislocation".

The basic difference between the earthquake and AE is the scales of cracking (faults) and the related frequency ranges. As shown in Fig. 2.5, the earthquake involves elastic waves of low-frequency components up to several Hertz (Hz), while the AE phenomenon involves the emission of waves of high-frequency components of up to several MHz. In general, humans can physically sense seismic events, while inaudible waves with low amplitudes are generated as AE phenomena.

In the case of the earthquake, seismic waves measured by a seismograph are analyzed to locate the hypocenter, and the scale of the earthquake is estimated as the seismic magnitude. Similarly, in the case of AE, the location of AE (AE source) and the scale of damage resulting in AE can be estimated by analyzing measured AE waves.

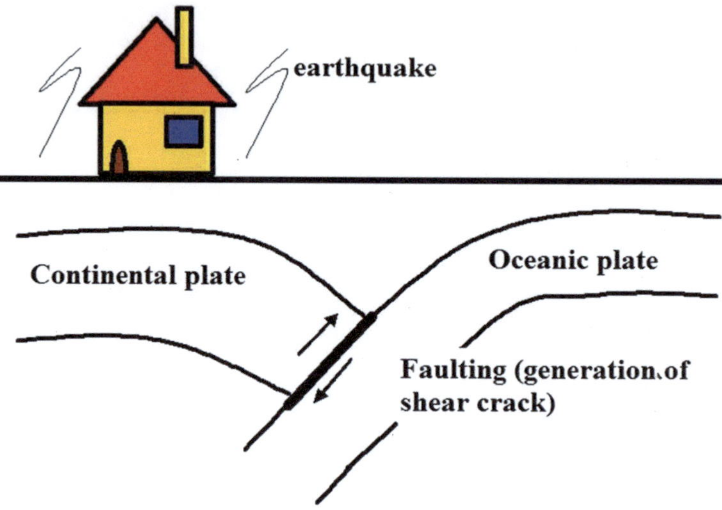

Fig. 2.3 Generation mechanism of a plate earthquake

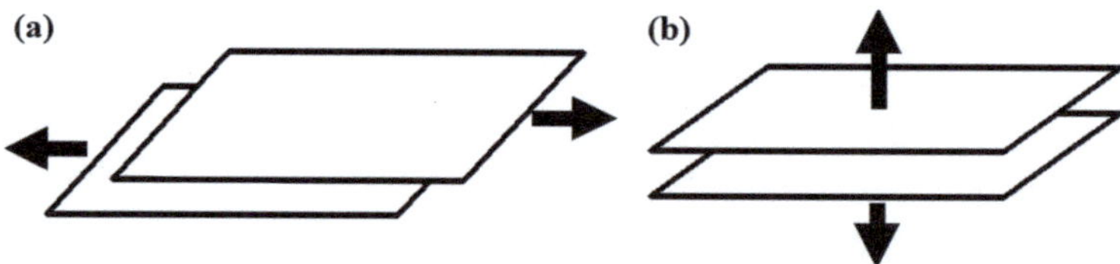

Fig. 2.4 Types of crack motions at a crack surface. (**a**) In-plane shear dislocation (slip crack-motion). (**b**) Tensil dislocation (opening crack-motion)

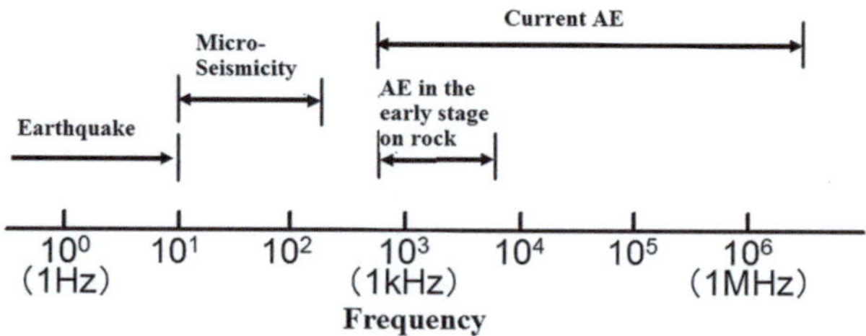

Fig. 2.5 Frequency ranges of various elastic waves

2.2 AE Testing as Non-destructive Testing

Masayasu Ohtsu

AE is used for NDT, regardless of the amount of strain energy released during the deformation and failure of materials. In this section, the reasons for choosing AE testing (acoustic emission testing, AT) and the cases in which AT is employed for NDT are described.

Failure, which does not occur instantly, generally begins at the micro-level, leading to final failure after the gradual accumulation of micro-level cracks. AT is a method of monitoring or measuring this process up to the final fracture. In other words, by detecting AE during the initial phases of the generation and growth of any crack (repairable phases), we can stop ongoing operation and test to avoid ultimate failure of the machine or structure (Fig. 2.6). This is why AT is used for NDT.

An ultrasonic testing (UT) is widely used as a means to detect cracks (flaws) in a material. UT and AT both involve the use of elastic waves and are often compared. Consequently, we describe the characteristics of UT and AT and the difference between the two.

In UTs (Fig. 2.7a), a reflected wave (echo) from a crack is detected by the excitation of an elastic wave from a probe, and thus, the crack is detected. In ATs (Fig. 2.7b), a crack is discovered by detecting an AE wave released from it. Both these methods involve the use of elastic waves, but there is an essential difference

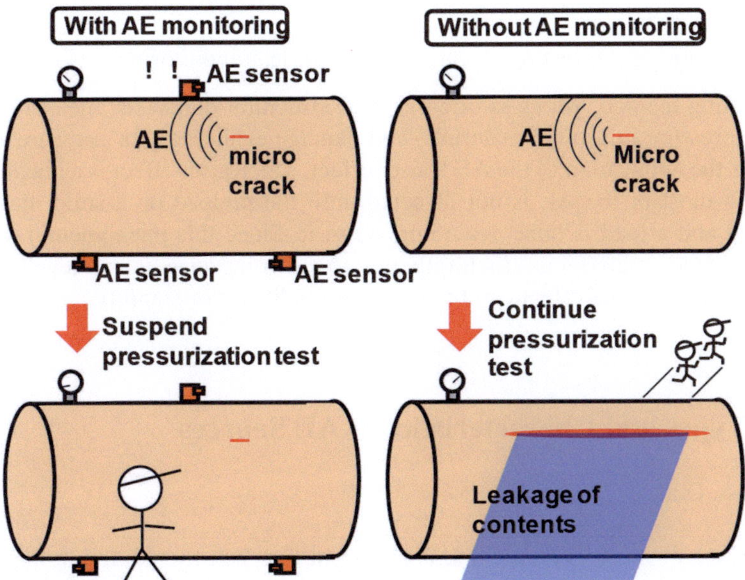

Fig. 2.6 Application to pressure vessels

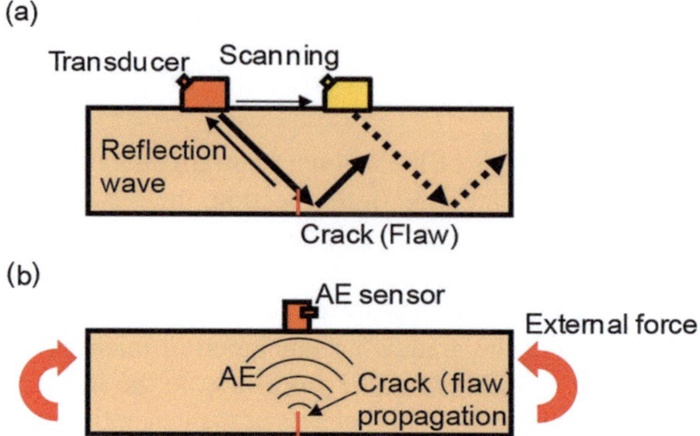

Fig. 2.7 Comparison between ultrasonic testing and acoustic emission testing. (**a**) Ultrasonics testing. (**b**) AE testing

between the two in this respect. Recently, research on an acoustic-ultrasonic (AU) method in which ATs and UTs are combined has been promoted.

As mentioned above, the application of AT to structural diagnosis led to the prediction of earthquakes and falling rocks and the establishment of the current monitoring technique. Consequently, the following capabilities are being incorporated in commercial products.

1. Detection of AE phenomena associated with micro-failure or indications of failure
2. Detection of AE indicating damage by continuously observing noise and vibrations
3. Application of AE testing to various types of leak monitoring

A testing method, which is often used in structural diagnosis, measurement of ground pressure in ground materials, and damage evaluation for pressure vessels, involves the application of the AE Kaiser effect. The Kaiser effect is a phenomenon in which most of the AE is not detected until the preload on a solid material is removed and a load is once again applied to it. Since this phenomenon does not occur when the material is structurally unstable (for instance, in the developmental stages of failure), the effect can be used as an indicator of stability.

2.3 Types and Characteristics of AE Sources

Masayasu Ohtsu

As a source of AE waves, a crack is shown in Fig. 2.2. In addition, a martensitic transformation and metal transformations such as tin cry are well known as AE sources. On rocky and flat terrain that is old enough to contain metals, collapse phenomena such as falling and talking rock are considered to be sources of AE waves. These phenomena, essentially caused by a failure phenomenon, can be said to result from crack formations.

Furthermore, AE waves are generated by rust formation and friction caused by rust. The generation of AE waves by friction is not essentially different from the occurrence of an earthquake caused by a fault slip. However, in the case of a composite material, delamination and fiber fracture occur in the material. Therefore, AE is caused by a complicated combination of these events.

Next, the detection of abnormal noise for acoustic diagnosis has been carried out to monitor the safety of equipment and determine when the equipment must be replaced. Abnormal noise caused by damage to tool edges and turbine blades can also be considered as AE sources. Consequently, the AE method is applicable to the detection of such noise. Furthermore, AE testing has already been employed in nuclear facilities to monitor loose parts of systems. This method detects the existence of broken pieces and fragments of parts (loose parts) that have slipped into the insides of pipes in a reactor.

Further, some of the materials reported so far do not generate AE upon their failure. This is considered to be due to the minimum energy of failure or ultra-high-speed failure. This does not mean that AE waves are not generated; rather, they cannot be as easily detected as in AE events.

AE waves are essentially generated by failure phenomena and can be mostly attributed to the formation of microcracks. This type of AE is sometimes called primary AE. On the other hand, AE generated by rust formation and friction caused by inclusions and particles is called secondary AE. Typical mechanisms that generate AE waves are conceptually illustrated in Fig. 2.8.

> ***Description of Term* (Transformation)**
> The micro-features of metal materials indicate that their atoms are arranged systematically (crystal structure). The crystal structure of a metal changes under certain conditions—this is called transformation.

Conventionally, types of AE waves are classified as burst AE (transient AE) and continuous AE waves. In principle, an AE wave is generated by the formation and growth of a crack at its source. Thus, a burst AE wave is reasonable. For instance, an elastic wave emitted by an opening crack propagates, causing amplitude damping. As a result, there is one waveform whose amplitude attenuates naturally over time. However, in the case of continuous and consecutive generations of dislocations, the AE waveform shown in Fig. 2.9 is observed, resulting from the overlapping effects.

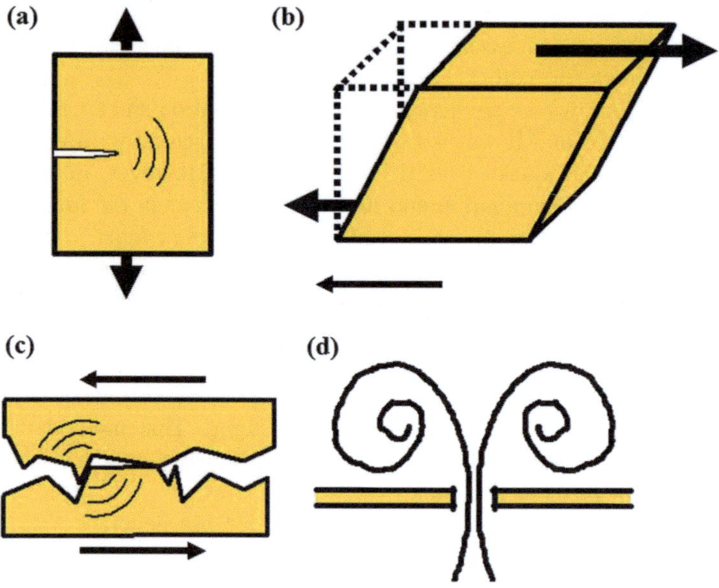

Fig. 2.8 Examples of sources of AE waves. (**a**) Cracking. (**b**) Deformation and transformation. (**c**) Sliding or slip. (**d**) Leakage

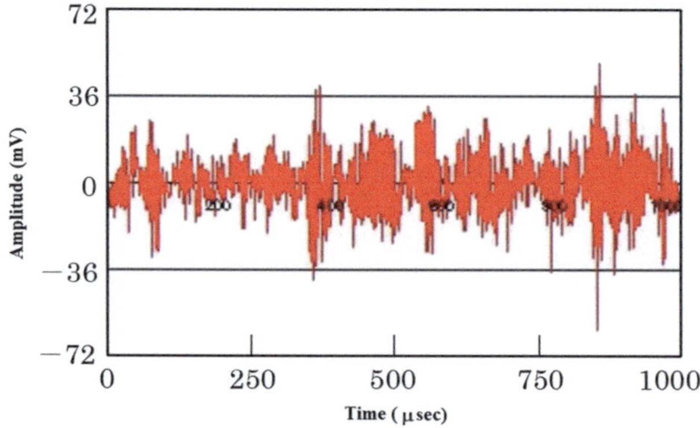

Fig. 2.9 Example of a detected AE waveform

As the time axis in the figure is extended, individual AE events can be discriminated in most cases. Consequently, they can be considered to be overlapping burst AE waves. In particular, AE waves associated with plastic deformation in metal materials are difficult to separate and are well known as continuous AE waves.

2.4 Properties of AE Waves

Masayasu Ohtsu

2.4.1 Propagation in Solids

An AE wave is, in principle, defined as an elastic wave generated at an AE source. Wave motion is a phenomenon in which particle motions are dynamically transferred to adjacent particles, as each particle only vibrates at its own position and does not move. However, wave motion itself propagates with its characteristic velocity.

Wave motions are defined on the basis of orientations of particle motions and directions of propagations. In an earthquake, for instance, the motion that people first feel is longitudinal, and the transverse motion arrives later. Major motion results from a Rayleigh wave, which is generated at the surface of the Earth after the longitudinal and transverse waves arrive. Similarly, AE waves consist of longitudinal, transverse and other waves.

1. Longitudinal wave (P-wave, Primary wave)

   In the case of longitudinal waves (Fig. 2.10a), particles vibrate along the direction in which the wave propagates. The wave is also called a dilatational wave since it is associated with volume change. The longitudinal wave can propagate through all media types—solids, liquids, and gases—and is the fastest of elastic waves.

2. Transverse wave (S-wave, Secondary wave)

   In the case of transverse waves (Fig. 2.10b), particles vibrate perpendicular (in the "lateral direction") to the direction in which the wave propagates. The transverse wave is also called a shear wave, and can propagate through solids but not through liquids and gases. The ratio of the velocity of the transverse wave to that of the longitudinal wave is theoretically equal to $\sqrt{\frac{1-2v}{2(1-v)}}$, where v is Poisson's ratio.

3. Other waves

   Longitudinal and transverse waves are always generated and propagate in an elastic solid. They are sometimes called body waves. Following their arrival at the surface, other waves are generated in a solid. For instance, Fig. 2.10c shows the propagation of Rayleigh wave. In the case of Rayleigh wave, particles near the surface move elliptically, and motion decreases with depth. It is thus called a surface wave. In a thin plate, Lamb waves (plate waves) are generated, as the plate vibrates. The Lamb wave has a symmetrical mode (S mode) in which the plate vibrates symmetrically, as shown in Fig. 2.10d, and an anti-symmetrical

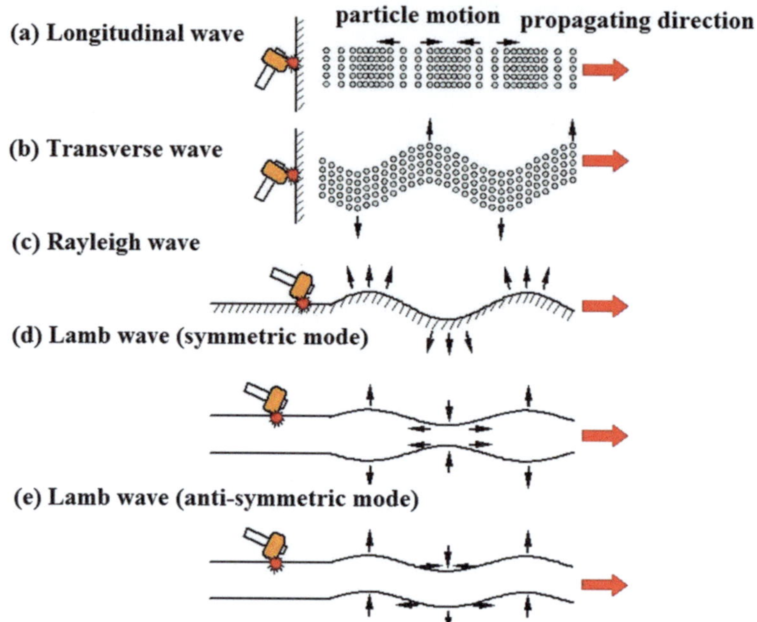

Fig. 2.10 Types of wave motion

mode (A mode) in which the plate vibrates asymmetrically, as shown in Fig. 2.10e.

2.4.2 Wave Velocity and Wavelength

The wave velocity at which an elastic wave propagates through a solid depends on properties of the materials and the wave type. For instance, the longitudinal wave propagates through aluminum at a velocity of approximately 6350 m/s, while the transverse wave propagates at approximately 3130 m/s. Even for the same type of wave, the velocity varies, depending on the material. For instance, the velocity of the longitudinal wave propagating through concrete is around 4000 m/s.

The velocity of the longitudinal wave, C_L [m/s], is expressed in terms of Young's modulus E, Poisson's ratio v and the density ρ of the material through which the wave propagates:

$$C_L = \sqrt{\frac{(1-v)E}{(1+v)(1-2v)\rho}}. \tag{2.1}$$

On the other hand, the velocity of the transverse wave, C_T [m/s], is expressed as

$$C_T = \sqrt{\frac{E}{2(1 + \nu)\rho}}. \tag{2.2}$$

The velocities of the longitudinal and transverse waves propagating through typical materials are listed in Table 2.1. The velocity of the Rayleigh wave is approximately 90 % of that of the transverse wave. The velocity of the Lamb wave varies with the plate thickness, frequency, and mode.

The wavelength is defined as the distance of one cycle in the sinusoidal motion of a wave. An example of wave motion with constant frequency is shown in Fig. 2.11. The time interval between successive motions of one cycle is defined as the period and expressed as T [s]. The number of cycles per second is defined as the frequency and expressed as f in units of Hertz (Hz), which is equivalent to cycles per second. Consequently, the period T and frequency f are related by

$$f = {}^{1}/_{T}. \tag{2.3}$$

In the case that the wave propagates with velocity C [m/s], the wavelength is derived from the wave frequency f. Thus, the wavelength is expressed as λ [m], and the relation among the velocity C, frequency f, and wavelength λ is

$$\lambda = {}^{C}/_{f}. \tag{2.4}$$

When motion is repeated at f cycles per second, a wave with wavelength λ travels a distance of $f\lambda$ per second, which is equal to the velocity C as given in Eq. 2.4. For a velocity C, a higher frequency f results in a reduced wavelength λ.

For instance, the velocity of the longitudinal wave is 4000 m/s in concrete. Therefore, the wavelength of the 100 kHz component is obtained as 4000 m/s/ 100 kHz = 40 mm. A frequency band lower than 100 kHz is often employed for AE testing in rock and concrete. Hence, it is noted that wavelengths of several centimeters are normally measured. As a result, scattering due to inclusions and aggregates becomes minor. In contrast, the velocity of the longitudinal wave is 5900 m/s in steel, and the wavelength of the 1 MHz component is obtained as 5900 m/s/ 1 MHz = 5.9 mm. Consequently, the wavelength often becomes greater than the plate thickness. In this case, dominant motions of AE waves result in Lamb waves, instead of longitudinal and traverse waves. Therefore, it is necessary to pay a particular attention to selecting a velocity for locating AE sources (Sect. 2.6, Chap. 2).

2.4.3 Attenuation

The attenuation occurs in time and in space. The former is related to time-series motion of a material and is generally called viscous damping. The latter results in damping of the amplitude with distance and is called distance attenuation.

Table 2.1 Velocities of typical materials (representative values)

Medium	Density (kg/m^3)	Velocity of longitudinal-wave (m/s)	Velocity of transverse-wave (m/s)
Aluminum	2700	6350	3130
Steel	7800	5900	3200
Concrete	2500	4000	2600
Water	1000	1430	–
Air	1.2	330	–

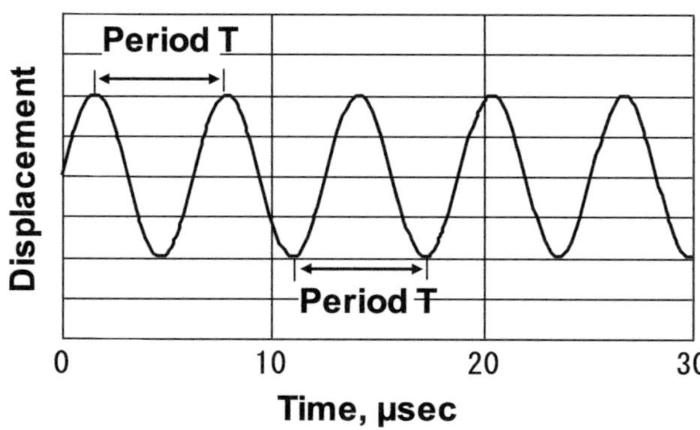

Fig. 2.11 Wave motion and the period

As an AE wave propagates from its source, its amplitude decreases owing to attenuation (Fig. 2.12). Hence, it becomes necessary to determine the proper positions and number of AE sensors, depending on sensor-to-sensor distances. Figure 2.13 shows a relation between detectable AE waves and frequencies in concrete. In general, AE waves with higher frequencies can attenuate significantly. Therefore, the sensors must be as well positioned so that the distance between the source and sensors is determined, taking into account detectable AE waves at object frequencies.

2.4.4 Reflection and Transmission

AE waves are generally detected by an AE sensor placed on the surface of a material. Therefore, wave components affected by reflection on the sensor-installed surface are measured. It should be noted that both longitudinal and transverse waves are generated by reflections of longitudinal waves. Even the incidence of a transverse wave generates a longitudinal wave through reflection. Snell's law is

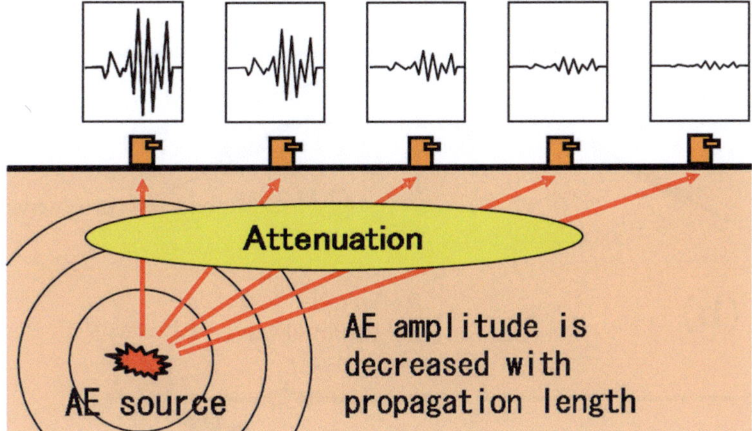

Fig. 2.12 Schematic illustration of attenuation

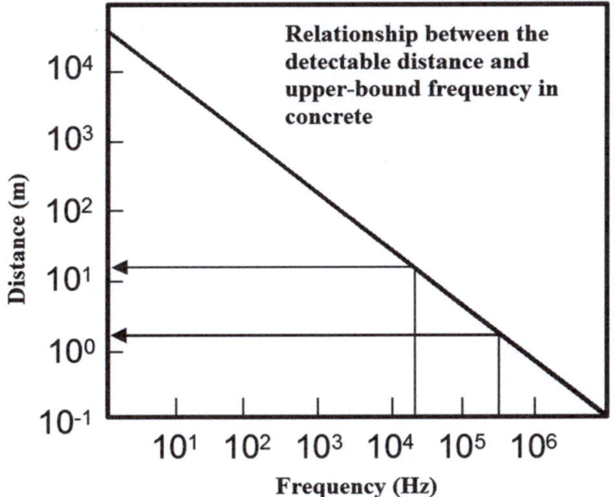

Fig. 2.13 Attenuation rate for concrete

known as a theory governing incident and reflection angles. This law is expressed as a relation between the propagation velocity C and an incident angle θ:

$$\frac{C_1}{\sin\theta_1} = \frac{C_2}{\sin\theta_2},$$

(2.5)

where θ_1 is the incident angle, C_1 is the propagation velocity, θ_2 is the reflection angle, and C_2 is the propagation velocity of the reflection. In the case of an incident transverse wave (S wave) as shown in Fig. 2.14a, we have

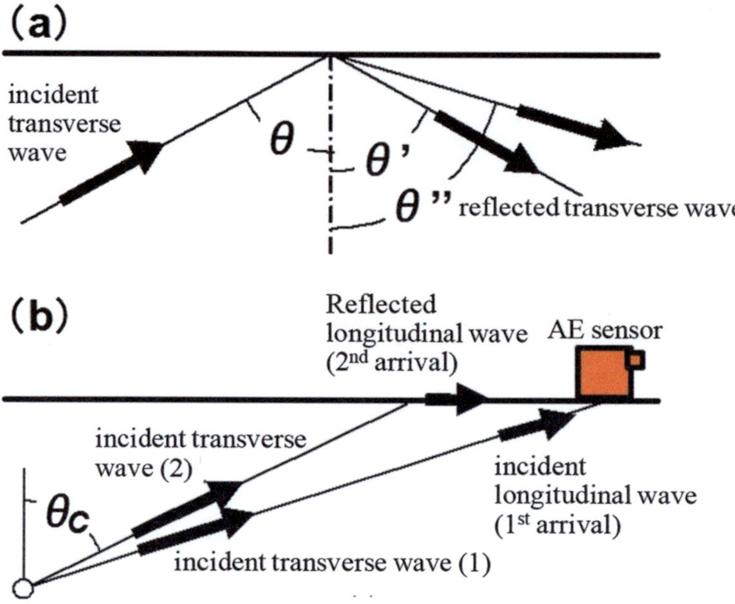

Fig. 2.14 Reflections of AE waves at the surface due to the incident of the transverse wave

$$\frac{C_T}{\sin\theta} = \frac{C_L}{\sin\theta'} = \frac{C_T}{\sin\theta''} \tag{2.6}$$

As a result, the incident angle θ becomes equal to the reflection angle θ'' since the velocities of the transverse waves, C_T, are equal even after reflection. Since the velocity of the longitudinal wave, C_L, is larger than that of the transverse wave, C_T, the reflection angle θ' becomes larger than θ. When the transverse wave (S wave) arrives at the AE sensor installed on the surface at a critical angle θ_c, the reflected longitudinal wave (P wave) propagates along the surface. Since this wave (SP wave) propagates faster than the direct transverse wave, AE waves are detected at the AE sensor as the incident longitudinal wave, the SP wave, and the transverse wave successively. These waves are followed by the Rayleigh wave.

Reflection can be considered to be a special case in which a boundary surface exists between two materials (propagation media) and no upper layers exist in transmission. In the two layers shown in Fig. 2.15, an incident wave becomes a reflected wave at the boundary surface and then becomes a wave transmitted to other layers. This transmitted wave causes refraction. An incident wave at the given incident angle enters other layers at a given angle of refraction. The relation between the incident and refractive angles is given by Eq. (2.6).

According to the theory of elastic wave motion, the amplitude of the incident wave and the ratio of the amplitudes of the reflected and transmitted waves in Fig. 2.15 are known to be related. Assuming that the amplitude of an incident wave

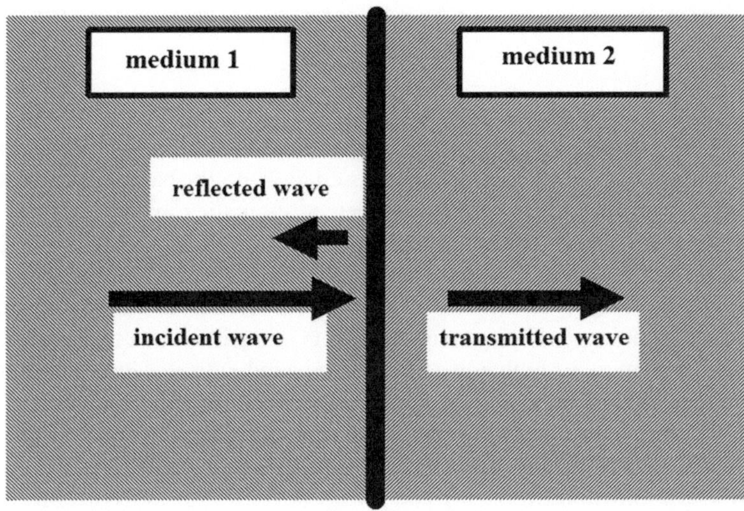

Fig. 2.15 Reflection and transmission of an AE wave

is A, that of a transmitted wave is A_T, and that of the reflected wave is A_R, these relations are

$$\text{Reflection coefficient} \quad \frac{A_R}{A} = \frac{\rho_2 C_2 - \rho_1 C_1}{\rho_2 C_2 + \rho_1 C_1}, \tag{2.7}$$

$$\text{Transmission coefficient} \quad \frac{A_T}{A} = \frac{2\rho_2 C_2}{\rho_2 C_2 + \rho_1 C_1}, \tag{2.8}$$

where ρ_1 is the density of Medium 1 in Fig. 2.15, C_1 is the propagation velocity of Medium 1, and ρ_2 and C_2 are the density and propagation velocity in Medium 2, respectively. In the absence of Medium 2, since $\rho_2 C_2 = 0$, the reflection coefficient A_R/A becomes -1, while the transmission coefficient A_T/A becomes zero. On the other hand, if the relation $\rho_2 C_2 = \rho_1 C_1$ is established in the same medium, no wave will be reflected with a reflection coefficient zero and all waves will be transmitted with a transmission coefficient of 1.

2.5 Signal Processing

Mitsuhiro Shigeishi

Detecting AE waves for inspection can be compared with estimating and comprehending the condition of machines used daily on the basis of their operation sounds. A skilled machine operator can detect anomalies in the machine on the basis of "unusual noises from the machine." This indicates that "something is wrong with the machine." An engineer familiar with the components of the machine can often discover the defect on the basis of unusual noises. For instance, intermittent low noise or rattling sounds indicate that parts are rattling and bolts are loose, while a continuous high noise or rubbing sound indicates the wear of rotating parts. The machine is diagnosed on the basis of noise. When people hear a sound, they can recognize the sound through its attributes such as its interval (tone) and tune (rhythm) and then assess the situation on the basis of the sound using their knowledge and experience.

However, to conduct a series of tasks using the machine (for instance, to detect a sound, identify its characteristics, and accurately evaluate the sound), it is required to operate the machine in many special processing tasks. AE involves the propagation of weak waves through a solid and no sound can be immediately heard by a person. Various devices are required for the detection of such waves and the identification of their characteristics on the basis of acoustic phenomena.

2.5.1 What Is an AE Signal?

In scientific as well as AE measurements, physical quantities including sound, vibration, temperature, and light intensity are called signals. These quantities must be observable and are generally converted to electric signals using an appropriate sensor.

An electric circuit needed for AE measurement is called an AE channel. The channel mainly consists of (a) an AE sensor, (b) a preamplifier or an impedance matching transformer, (c) a filter, (d) a main amplifier or other necessary devices, (e) a cable, (f) a detector or processor, or a combination of devices that have the same functions as these devices (see Fig. 2.16). In other words, an AE signal obtained from AE measurements is a physical quantity determined after an AE wave propagating through a solid is detected by these devices and converted to an electric signal. The quantity is generally a voltage value [mV].

Next, the information included in the AE signal is described. AE is a phenomenon in which some of the energy released by local changes in a material (an AE event) propagates through the material as a wave. Consequently, the original AE signal must contain information on the nature of an AE event or the source of the wave. A waveform of the type of AE signal shown in Fig. 2.17 is generally called a

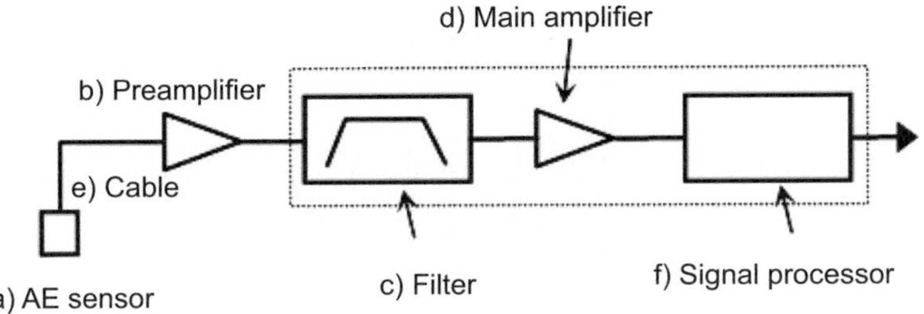

Fig. 2.16 Basic AE channel

burst waveform. As apparent from the name, any burst change in equilibrium generates AE. In addition, on the basis of the scale of the wave (height in the figure), the degree of the change can be estimated. For instance, assuming that the change is the burst breakage of given parts, the degree of the breakage or the size of the broken parts can be inferred from analogy. If this breakage results from the collision of a flying object with an object, the information required to determine the degree of damage to the object is included in the AE signal.

However, until the wave propagates from the AE source to a point where it can be detected or a point where an AE sensor is present, depending on the property of the solid in which the propagation occurs, some or all of the information on the source contained in the wave may be lost or information on the material properties may be gained. Consequently, the information must be carefully collected.

2.5.2   Basics of Waveform Parameters

In the case of material evaluation and structure diagnosis based on AE, the relation between an AE source event and the waveform characteristics in the form of the graphical representation of the AE signal is generally understood.

(a)   a burst waveform with clearly observable longitudinal and lateral waves;
(b)   a harmonic continuous waveform with longitudinal and lateral waves that are difficult to observe;
(c)   a burst waveform with longitudinal and lateral waves that are slightly difficult to observe.

Waveforms obtained for the AE signal vary widely but can often be visually classified into certain patterns. To illustrate this, three waveforms are shown in Fig. 2.18. From the figure, a difference in the wave intensity, duration of motion or type of propagation can be observed. In this manner, the characteristics of a waveform are clearly expressed as values such as motion intensity, wave-continuation time, and the time from the start of wave generation to the time

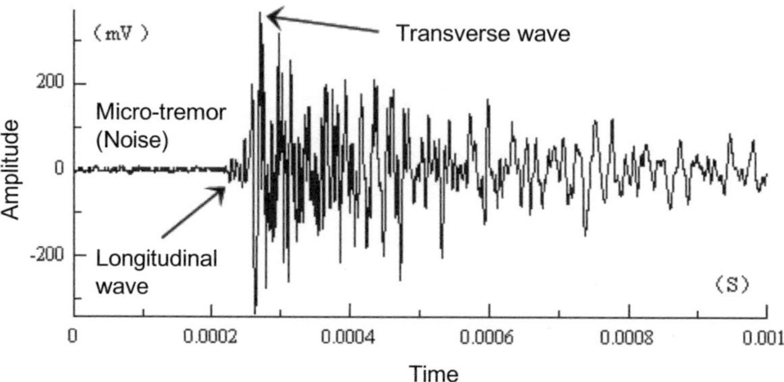

Fig. 2.17 Example of an AE signal

when the largest wave is observed or the time from the beginning of the reduction of wave intensity to the extinction of the wave, which are called waveform parameters. Consequently the characteristics of the AE signal are described by these waveform parameters.

2.5.3 Discrimination of an AE Signal

A wave generated in one AE event comprises a block of various wave components such as longitudinal, transverse, or surface waves, as shown in Fig. 2.17. Because of the difference in the propagation velocities of different wave components, these components successively reach the AE sensor placed far from the AE source with a time lag. Furthermore, there is a time lag between the gradual reduction in wave intensity to the extinction of these waves. A series of pulses is called a wave packet. Furthermore, the wave packets of the AE signal corresponding to the AE waves released from a certain source are separated and extracted from signals that are continuously output by the AE sensor. This signal processing is called discrimination.

Actual AE signals, in contrast to deterministic signals with a regular time period and amplitude, are random signals in which the changes in the signal values after a certain time instant cannot be predicted. In addition, as shown in Fig. 2.18, because several types of noises generated by many factors are included in AE signals to be measured, it is unclear what the start and end of AE signals are. Therefore, instruments for AE measurements rely on unique methods to differentiate between AE signals.

As shown in Fig. 2.19, when a voltage threshold is set slightly higher than the voltage level of the background noise during AE measurements and the amplitude of the AE signal exceeds this threshold, the electric signal is recognized as an AE signal. In addition, this voltage threshold can be set at the dashed line for which the

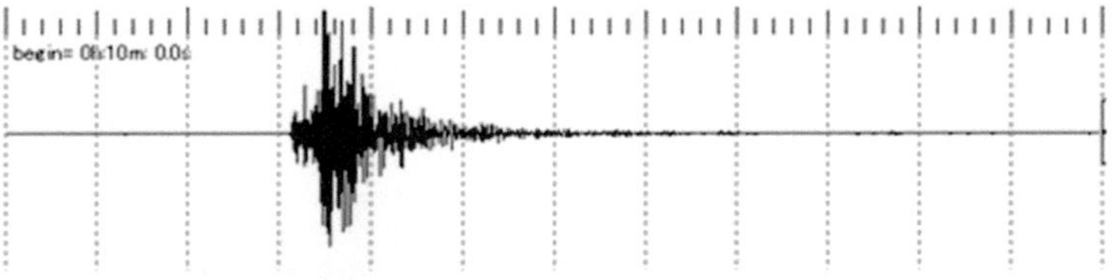

(a) Burst waveform with clear arrival of longitudinal and lateral waves

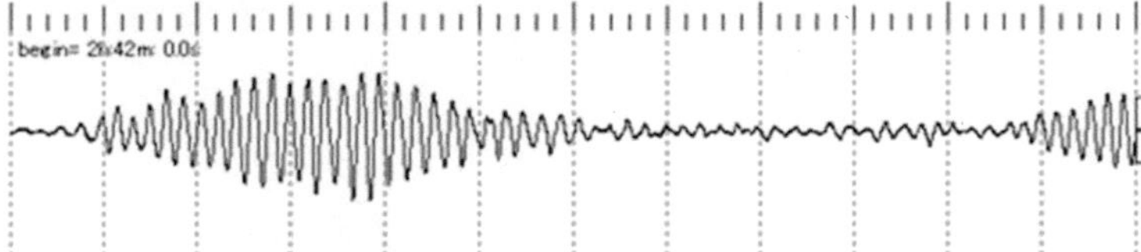

(b) Harmonic continuous waveform mixed with longitudinal and lateral waves

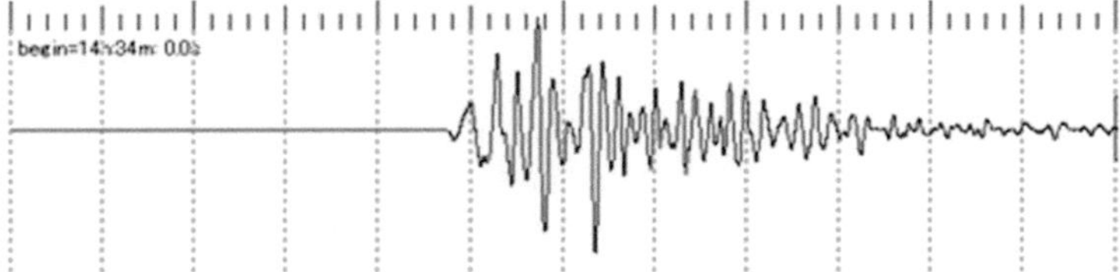

(c) Burst waveform with unclear longitudinal and lateral waves

Fig. 2.18 Classified waveforms of AE signals

voltage is negative in Fig. 2.19. Furthermore, these two methods can be concurrently used.

However, in this simple processing, the discriminated AE signals form a pulse train, resulting in potential loss of critical information on AE. Consequently, in this case, a method illustrated in Fig. 2.20 is sometimes used for signal processing. In particular, the amplitude of an AE signal is measured while an envelope is detected from the AE signal. The start of the AE signal is determined to be the point at which the voltage level of the detection signal first exceeds a set voltage threshold. On the other hand, the end of the AE signal is determined to be the point at which the voltage level of the envelope detection signal falls below a set voltage threshold. One wave packet of continuous AE signals between the start and end of the AE signal is called an AE hit.

Fig. 2.19 Signal discrimination with a voltage threshold

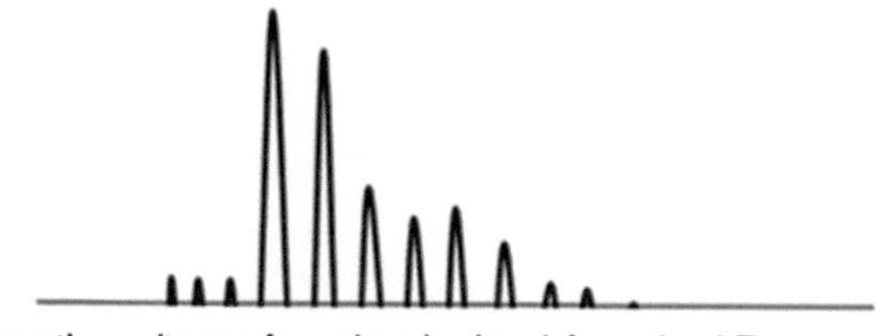

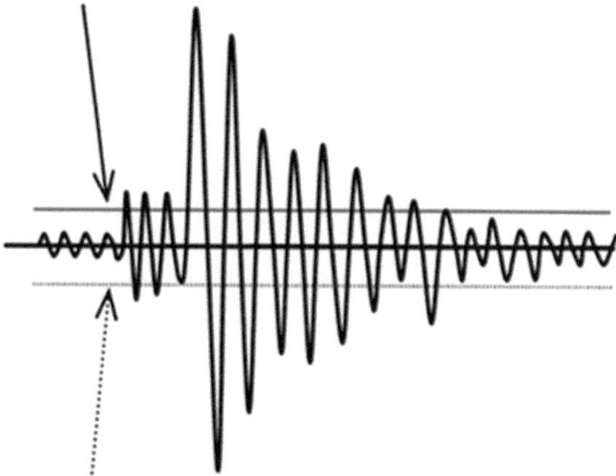

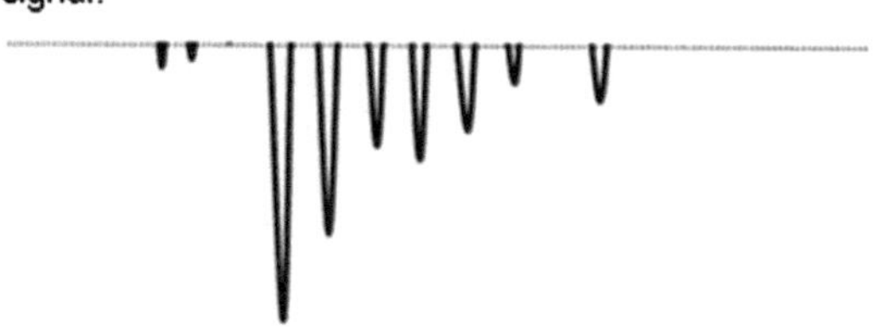

Furthermore, there is another method for the concurrent use of this voltage threshold and time discrimination by a timer (see Fig. 2.21). In particular, assuming that the start of the AE signal is the point at which the voltage level of the AE signal first exceeds the voltage threshold, the AE signal levels and the voltage thresholds that are continuously input are compared. If the conditions for the voltage of an AE signal to be higher than a certain threshold are satisfied within a given time, the above comparison is repeated. If the conditions cannot be satisfied even when the time has elapsed, this point is determined to be the end of the AE signal.

It is considered that the discrimination of AE signals using the above mechanism will be easiest when the steps below are followed.

1. Set a voltage threshold Vt and detection time limit Tt that specify the start of the AE signal.

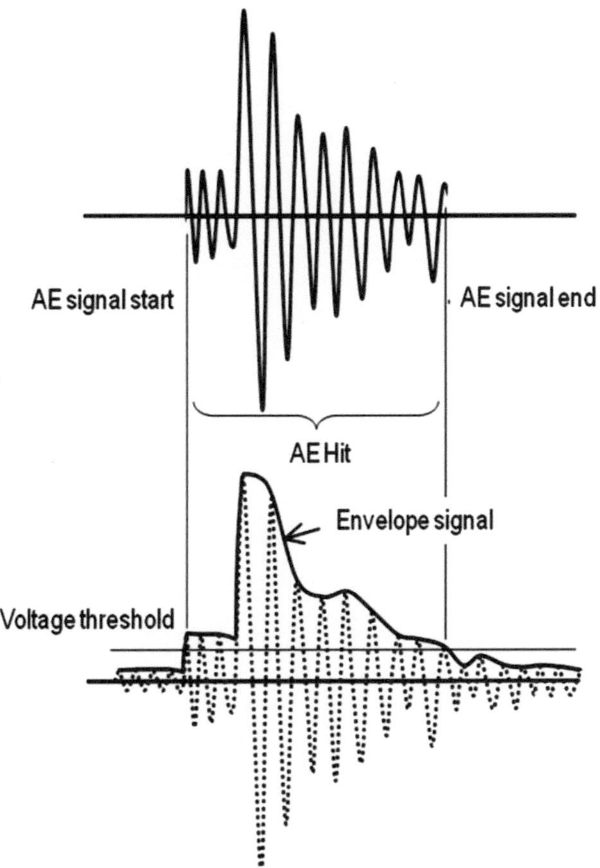

Fig. 2.20 Signal discrimination by envelope detection

2. Assume the point at which the level of the AE signal, *Vae*, first exceeds *Vt* as the start of the AE signal (AE signal start), and start the timer.
3. If *Vae* exceeds *Vt* again before the timer has reached the detection time limit *Tt*, reset the timer at this point.
4. Repeat the comparison between *Vae* and *Vt* until the timer reaches *Tt*.
5. If the timer reaches *Tt* without *Vae* exceeding *Vt*, consider this point as the end of the AE signal end.

Regardless of the discrimination processing of any AE signal, a given "dead time" during which the AE signal is not detected after the end of the AE signal is generally set. This time is established so that a wave packet of an AE signal already identified is not detected as a wave packet of another AE signal by the same AE sensor because of bypassing and reflection.

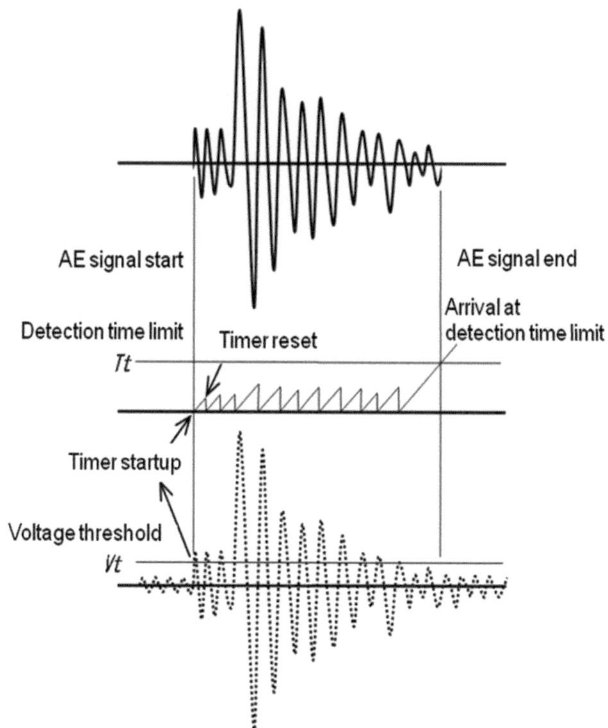

Fig. 2.21 Signal discrimination by timing parameters

2.5.4 AE Parameters

It was mentioned earlier that an discriminated AE signal could include information on an AE event in the source generating the wave. Therefore, information on AE events and physical phenomena is generally inferred indirectly; i.e., by calculating waveform parameters representing the characteristics of a waveform of the discriminated AE signal, examining temporal variations in specific waveform parameters during measuring AE, obtaining distributions for one or more predetermined evaluation thresholds, and investigating the correlation between different waveform parameters.

The focus of this section is the waveform parameters defined in terms of the AE standard, ISO 12716: 2001. Figure 2.22 illustrates the significance of the main parameters through schematic waveforms of an AE signal.

In general, the voltage threshold is only focusing on the instantaneous amplitude of AE. Though, evaluation threshold which is different from voltage threshold is used in some cases when calculating AE parameters. In Fig. 2.22, the evaluation threshold has the same value as the voltage threshold. In some cases, the evaluation thresholds are set to higher than voltage thresholds to calculate the AE parameters.

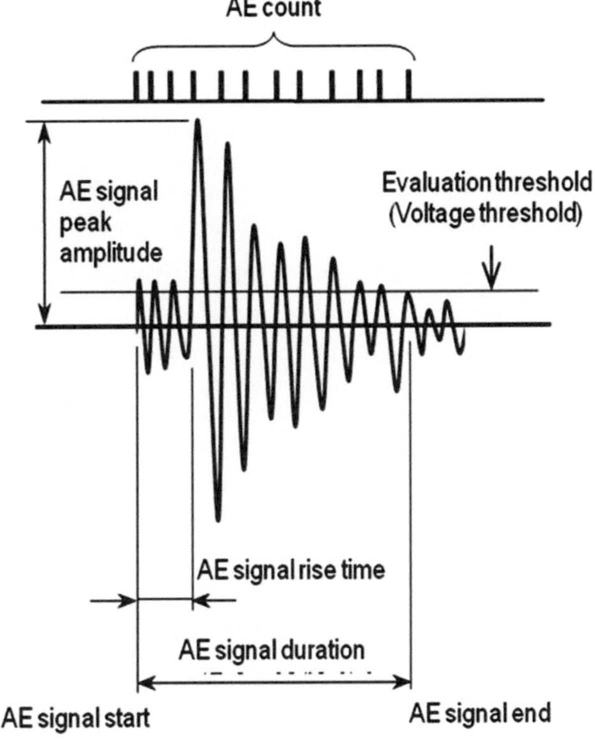

Fig. 2.22 Waveform parameters

The time that elapses between the start and end of an AE signal is called the AE signal duration, and the peak amplitude of the AE signal is called the AE signal peak amplitude or simply the AE signal amplitude. The time that elapses between the start of an AE signal and the time corresponding to the AE signal peak amplitude is called the AE signal rise time.

The frequency at which the AE signal amplitude exceeds a predetermined threshold within a specified time range is called the AE count, count, ring-down count, or emission count. In general, the AE count is related to the frequency at which the AE signal amplitude exceeds a predetermined evaluation threshold within the AE signal duration. The AE count is used in AE testing to determine the number and frequency of AE signals.

The AE count rate or emission rate is calculated from the AE count per unit time, as a parameter that indicates the increase/decrease in the frequency associated with the AE count time.

It is important to understand how AE signal processed in the AE measurement instrument that is currently in use, since waveforms and AE parameters may be changed when inappropriate frequency filter is used. It is also noted that evaluation threshold also affects AE parameters.

Comparisons of individual measurement results carried out by repeating the same AE measurement and continuous or intermittent AE measurements of the same object over a relatively long time would not be valid if the same measurement instrument is not used.

The purposes of AE tests are to quantify the characteristics of waveforms of AE signals using various methods and to parameterize them, to guess the (unknown) events that have occurred, determine the number and frequency of AE signals, and estimate when AE occurred. AE parameters are summarized as follows.

(a) Parameters based on the signal waveform level

1. AE signal (peak) amplitude

The maximum voltage in an AE signal (hereafter, referred to as the discriminated signal) detected between the start and end of the AE signal discriminated on the basis of a given measured threshold; the signal belongs to a series of AE signals generated by one AE event.

The voltage value of the AE signal can be expressed as a common logarithm on the basis of a reference value of 1 μV.

$$AE\ signal\ peak\ amplitude\ [dB_{AE}]\ =\ 20\ log_{10}(A_1/A_0)$$

Here, $A_0 = 1$ μV, which is the output from the AE sensor without any amplification, and A_1 is the measured maximum voltage of the AE signal.

2. AE count (ring-down count)

The frequency when the measured voltage of the AE discriminated signal exceeds an evaluation threshold. The AE count at a given time is divided by the time that has elapsed since the start of the AE signal, and the count result per unit time is called the AE count rate.

3. AE energy

The AE signal energy is determined as the square of the AE signal amplitude, the integral of the square of the instantaneous amplitude of the AE discriminated signal over a certain duration, or the integral of the instantaneous AE signal amplitude determined by envelope detection over a certain duration. However, these values differ from the total energy (AE event energy) released due to an AE event.

(b) Parameters based on the shape of the signal waveform

4. AE signal duration

Time that has elapsed between the start and end of the AE discriminated signal.

5. AE signal rise time

Time that has elapsed between the start of the AE discriminated signal and the time at which the AE signal peak amplitude is attained.

6. Ratio of rise time to amplitude

The ratio of the AE signal peak amplitude to the AE signal rise time.

(c) Parameters based on the level of the continuous AE signal

    7. AE root-mean-square (RMS) value

The effective value (square root) is obtained by calculating the root-mean-square of an AE signal. The effective value is also called the root-mean-square value. The effective value of the AE signal represents the energy level of the AE signal and also enables us to evaluate the rate of occurrence of AE in the same manner as the AE count rate does, particularly in the case of continuous AE (refer to the description on page 32).

    8. Average signal level (ASL) of the AE signal

The average signal level represents the average energy calculated by integrating the absolute value of the amplitude of a rectified AE signal over a given time and dividing the result by the time that has elapsed since the start of the signal.

The average level of the AE signal is considered to be also effective for evaluating the root-mean-square value of a continuous AE signal (refer to the description in Chap. 4)

(d) Other AE parameters

AE hit time
AE hit count
AE hit count rate
AE event count
AE event count rate
Damping factor based on propagation distance
Average frequency
Frequency spectrum
Energy moment

2.6 AE Source Location

Manabu Enoki

2.6.1 One-Dimensional and Two-Dimensional Location Methods

A great advantage of the AE method is that the location of an AE source can be determined relatively easily. In contrast to the case for a UT, it is not necessary to scan the whole of an object using a probe when determining the position of damage to the object. The location of the damage can be measured using a fixed sensor. This method is the same as the location method for determining an earthquake center. In the location method, the wave velocity of the object and the arrival time of the AE wave must be known. Consequently, since a clear rising point of the waveform is observed in the detected burst AE waveforms generated by micro-cracking, it is easy to detect the arrival time of the AE wave, thereby enabling relatively accurate location. On the other hand, since a clear rising point in the continuous AE waveforms associated with plastic deformations is not observed, it is difficult to determine the arrival time of the AE wave. Therefore, an accurate location cannot be expected.

It is necessary to determine beforehand the velocity of AE wave to identify the location of damage to the object using AE wave. Since the first-arriving longitudinal wave can often be clearly detected when the object is sufficiently thick, only the velocity of the longitudinal wave is used. Because the longitudinal waves often cannot be clearly observed in the case of a thin-plate object, it would be better to use the velocity of Rayleigh wave or Lamb wave (refer to Fig. 2.10); these velocities can be clearly observed. Before AE measurement, it is necessary to check whether the correct location can be found using a simulated AE source (pencil lead break and pulse generator) after installing an AE sensor.

In general, it is desirable to find the three-dimensional location in determining the damage position. However, when the number of measurable channels and the number of sensors are limited or when the shapes of the object are specific, it is difficult to determine the three-dimensional location of the damage. With a known velocity, at least an arrival time difference is needed only for the order of a location to be determined; i.e., the minimum numbers of sensors required for one-dimensional location, two-dimensional location, and three-dimensional location are two, three, and four, respectively.

A method for determining the arrival time of the AE wave is important. When a clear rising point in the AE waveform is first observed, this point can only be the arrival time, but the rising point generally cannot be clearly identified in most cases because of noise. In this case, it is practical to calculate the arrival time assuming that it is given by the point exceeding a threshold or the point at which the waveform attains the initial peak. However, it would be also necessary to evaluate the precision of location achieved using the simulated AE source.

One-dimensional location in which a burst waveform is detected is now discussed. Assuming that there is AE between two AE sensors, the location of an AE source is x, and the locations of the two sensors are x_1 and x_2 (See Fig. 2.23), we derive an equation for t_1 and t_2 when the AE wave reaches the sensors:

$$Ct_1 = |x - x_1|, \quad Ct_2 = |x - x_2|, \tag{2.9}$$

where C is the velocity of the AE wave. Furthermore, assuming that there is a difference between the arrival times at the two sensors (arrival time difference: Δt_{12}), we have

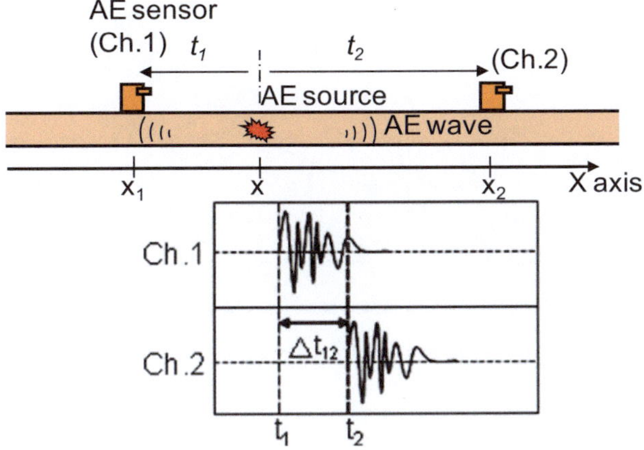

Fig. 2.23 Linear source location

$$C\Delta t_{12} = |x - x_1| - |x - x_2|. \tag{2.10}$$

The velocity and the positions of the sensors are known. From this equation, the one-dimensional location of the AE source, x, can be determined.

Similarly, three sensors enable us to determine the planar source location or the two-dimensional source location (x, y). Assuming that the velocity is C, the positions of the three sensors are (x_1, y_1), (x_2, y_2), and (x_3, y_3), the arrival times at the sensors are t_1 and t_2, and the differences between the arrival times at the sensors are Δt_{12} and Δt_{13}, as mentioned above. We thus have (Fig. 2.24)

$$
\begin{aligned}
C\Delta t_{12} &= \sqrt{(x - x_1)^2 + (y - y_1)^2} - \sqrt{(x - x_2)^2 + (y - y_2)^2}, \\
C\Delta t_{13} &= \sqrt{(x - x_1)^2 + (y - y_1)^2} - \sqrt{(x - x_3)^2 + (y - y_3)^2}.
\end{aligned}
\tag{2.11}
$$

These two equations with two unknowns x and y can be solved. The numerical solution of non-linear equations obtained using a computer enables us to easily determine the two-dimensional location of the AE source. The precision of location achieved by these methods depends mainly on the resolution of the arrival times, sensor sizes, and the sensor-to-senor interval.

2.6.2 Guard Sensor

As described in the previous section, if a sufficient number of measurement channels are available for measuring the arrival time of an AE wave, the location of an AE source can be determined. However, when many extraneous noise signals are measured from sections other than the object to be measured, it is inefficient to conduct location in the presence of assumed noise signals with the above method.

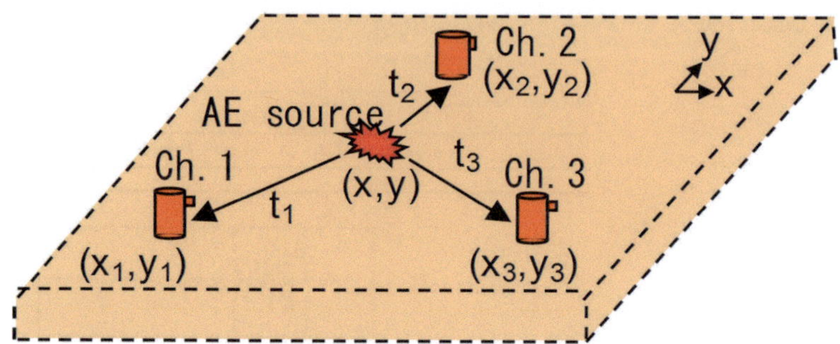

Fig. 2.24 Two-dimensional (planar) source location

The following method is useful for eliminating noise from the chucking of a test specimen, as required in a materials tensile test, or for removing noise generated by the test. It also contributes to the elimination of clear noise propagating from outside the object, even in a large structure.

The measurement method involving the use of a guard sensor for one-dimensional location is described below. As shown in Fig. 2.25, an AE source between the sensors at x_1 and x_2 is to be monitored. Guard sensors for noise elimination are installed at x_0 and x_3 outside the monitored area. If any noise is generated outside the area bounded by x_0 and x_3, the first signal will always reach a guard sensor at x_0 or x_3. Consequently, when comparing the arrival times of all AE signals in the channel, if a guard sensor at x_0 or x_3 indicates the arrival time of the first signal, then the noise in this event can be concluded as originating from outside the monitored area. Thus, the one-dimensional location of events except eliminated events can be efficiently carried out using the sensors at x_1 and x_2.

2.6.3 Zone Location

The concept of the first-hit channel described in the previous subsection is also useful in zone location, particularly in the case of large structures. The frequency band used in AE measurement is lower than that used in a UT; further, the signal damping in AE measurement is less than that in the UT. Therefore, AE measurement can be used to monitor large structures. However, in the case of a larger object or a weaker generated signal, all AE signals cannot be recorded in all measured channels because of signal attenuation. In this case, it is impossible to locate the AE source when there is a difference between the arrival times of AE signals. The AE signal that arrives first corresponds to the AE source nearest a sensor in the channel; this enables approximate location of the AE source.

A method for two-dimensional zone location of the object is described below. As shown in Fig. 2.26, sensors are equally spaced in measuring AE.

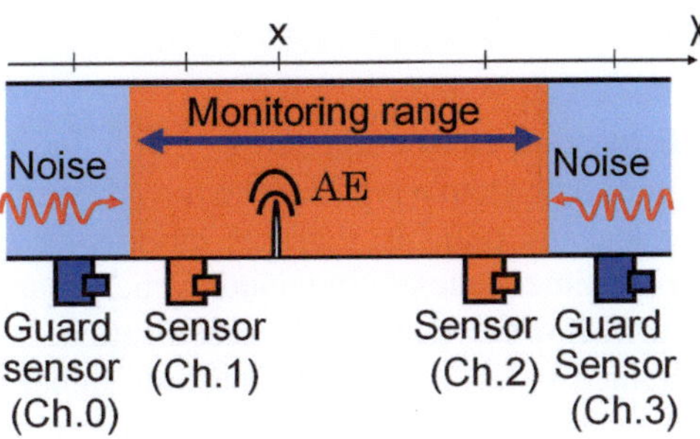

Fig. 2.25 Arrangement of a guard sensor for noise elimination

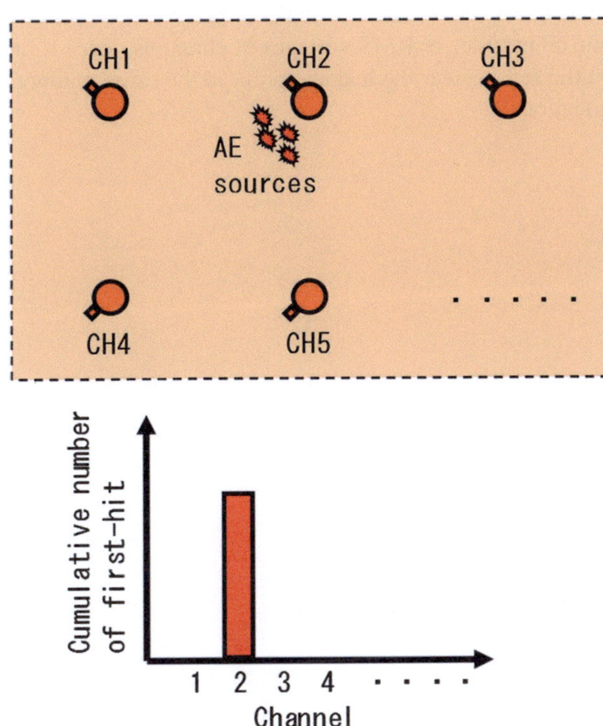

Fig. 2.26 Zone location

If any AE signal is detected, it can be concluded that an AE source exists near the sensor. For instance, if a histogram showing the number of events observed in each channel is plotted, the area with many AE sources can be identified. Consequently, the precision of locating AE sources will depend on the sensor-to-sensor intervals. Although this precision cannot be compared with that obtained from the difference in arrival times, this method significantly helps determine the damage to a large structure in a realistic manner.

2.6.4 Leak Location

AE location can be applied to determine the leakage points in various pipes. If liquids leak from a pipe and a burst AE signal is detected, it is possible to determine the leakage point in the pipe employing the above location method. When the arrival time is clearly obtained in each channel, as described earlier, the difference in arrival times enables accurate location. In addition, when clear signals cannot be obtained in all channels because of signal attenuation, zone location enables the determination of any leakage point.

On the other hand, gas leakage may generate a continuous rather than burst AE signal in most cases. Therefore, location employing the time axis is impossible in such cases. However, any leakage point can be roughly determined by comparing the amplitudes or RMS voltages of channels, since a strong signal must be obtained at the sensor near the leakage point in the same manner as in the above case of zone location.

Chapter 3
AE Sensor (AE Transducer)

Hidehiro Inaba

Abstract In this chapter, an overview of AE sensor (transducer) is presented. Conversion principal by the AE sensor is discussed. Structures and characteristics of resonant and broad-band AE sensors are explained. A calibration method of AE sensors is briefly introduced.

Keywords AE sensor • Conversion principal • Calibration method

3.1 Conversion Mechanism

Once AE is generated in solid materials, AE waves propagate within the solid materials and/or on the surface of the solid materials as elastic waves. During AE testing, these AE waves are detected by AE sensors attached to the surface of the solid materials. The AE sensor detects an AE wave and then converts it into a voltage signal. In general, this conversion is conducted using piezoelectric ceramics. In this section, the principle of this conversion mechanism by an AE sensor is explained.

Piezoelectric ceramic is one of the piezoelectrics. The basic physical behavior of piezoelectrics is illustrated in Fig. 3.1.

Electrodes are placed on both ends of a cylindrical piezoelectric element. When a voltage is loaded on a piezoelectric element via the electrodes, the piezoelectric element expands or contracts in accordance with the voltage application. Conversely, the piezoelectric element will generate a voltage when its shape is altered by a force. The piezoelectric element has the ability to convert mechanical energy into electrical energy reciprocally. This is referred to as piezoelectric effect and the materials showing above property as a piezoelectric material. There are two kinds of piezoelectric materials. One is quartz without any artificial modification. The other one is piezoelectric ceramics made by burning several raw materials at high

H. Inaba (✉)
Fuji Ceramics Corporation, Fujinomiya, Japan
e-mail: inaba_h@fujicera.co.jp

The Japanese Society for Non-Destructive Inspection, *Practical Acoustic Emission Testing*, DOI 10.1007/978-4-431-55072-3_3

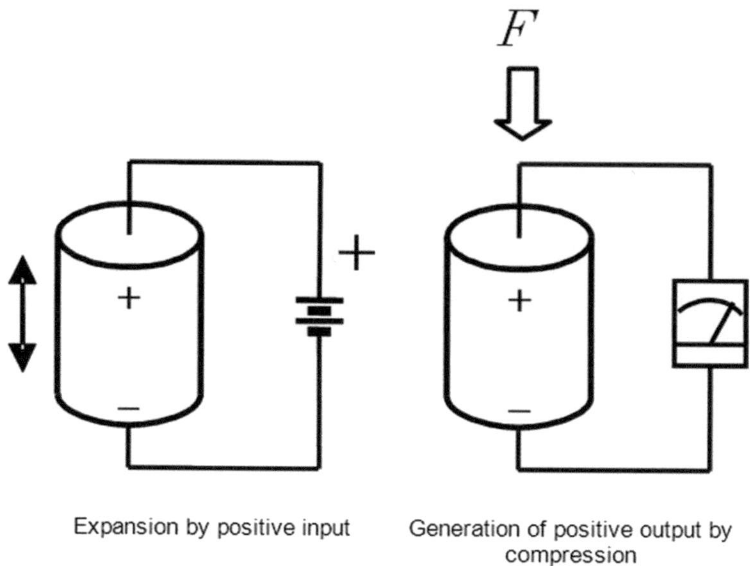

Fig. 3.1 Basic physical behavior of piezoelectric ceramics

temperature after a forming process. Piezoelectric ceramics are much better at generating electricity through deformation than other materials. Accordingly, piezoelectric ceramics are suitable for application in an AE sensor to achieve high sensitivity in measuring microscopical deformations, such as those of an AE waves. Although there are many types of piezoelectric ceramics, lead zirconate titanate (Pb (Zr, Ti)O$_3$), which is referred to as PZT, is widely used for AE sensors. Unlike quartz, PZT does not demonstrate piezoelectricity only by the forming and burning processes, but it shows piezoelectricity after a high voltage is applied (called polarization).

An AE wave propagates within solid material and/or along the surface of the solid materials, and can pass into an AE sensor fixed to the surface of the solid materials via the AE sensor's detection face. Since the detection face of an AE sensor is in contact with the solid surface and the inside surface of the detection face is adhered to a piezoelectric ceramic, the AE sensor can detect an AE wave with high sensitivity. When an AE wave reaches the piezoelectric ceramic after passing through the detection face, the piezoelectric ceramic is deformed by the AE wave. A voltage is then generated as a result of deformation of the piezoelectric ceramic. This voltage is called the AE signal and it passes from the AE sensor through a cable and preamplifier and is detected by measurement instrument.

3.2   Types and Structures of AE Sensors

3.2.1   *Resonance and Broad-Band Sensors*

AE sensors can be mainly classified as either resonance models or broad-bandwidth models. The schematic structures of these AE sensors are shown in Fig. 3.2.

First, the widely used resonance-type AE sensor is explained. When an AE wave reaches a piezoelectric ceramic through the detection face, the AE wave is repeatedly reflected within the piezoelectric ceramic (i.e., transfer element). During this reflection, the AE wave with the resonance frequency is emphasized and remains within the transfer element. In contrast, other components are attenuated quickly within the transfer element. Accordingly, the AE sensor achieves high sensitivity by taking advantage of the resonance provided by the transfer element. f_r, the resonance frequency of the transfer element depends on the thickness and sound velocity of the transfer element. Additionally, f_r is expressed by the following equation, where l and C are the thickness and velocity of the transfer element respectively. The value of f_r in this equation is the resonance frequency under the condition that the AE sensor is attached to the measuring object.

$$f_r = \frac{C}{4l} \tag{3.1}$$

In the equation, the value of f_r depends on the thickness of the transfer element, while in reality, the value of f_r is also affected by wave propagation in another direction such as the radial or width direction of the transfer element. Finally, overall f_r is characterized by the mutual influence among various kinds of f_r. Consequently, the sensitivity–frequency characteristics of the AE sensor are determined by the shape of the transfer element. In the case that we design a transfer element with lower resonance frequency, the element will be larger than those of higher frequency. In general, the size of the AE sensor will become large when f_r is below 60 kHz, while small when f_r is high.

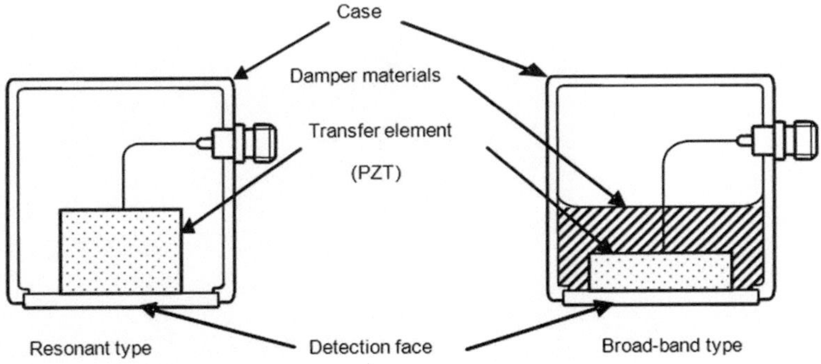

Fig. 3.2 Schematic structures of resonant-type and broad-band type AE sensors

A broadband AE sensor having a flat frequency response is used to confirm the frequency component and/or waveform analysis of AE waves. To acquire this flat frequency response, it is necessary to damp f_r in contrast with the case for the resonant type AE sensor. In terms of the structure, a transfer element is covered by a damper. The purpose of the damper is that an AE wave can pass from a transfer element to a damper without there being reflection of the AE wave in the marginal zone between the transfer element and damper, which suppresses the volume of the AE wave reflection and acquiring a flat frequency response.

3.2.2 Structure

As an AE sensor can detect very weak signals, its transfer element is normally installed within a metal case to shield the signal from outside noise. AE sensors are of various size; e.g., there is a minute sensor having a diameter and height of 3 mm, there is the common type having a diameter of 20 mm and height of 20–25 mm, and there are sensors for civil engineering with a diameter of 30 mm and height of 50 mm height with f_r of 30–60 kHz.

3.3 Characteristics of AE Sensors

Figure 3.3 shows a common resonance-type 150 kHz AE sensor. Figure 3.4 shows an example of the sensitivity–frequency characteristics of the sensor. The 150 kHz AE sensor is common in that it has a diameter of 20 mm and height of 23 mm. The alumina (Al_2O_3) detection face is white and located in the center of the top face.

Fig. 3.3 150-kHz resonant-type AE sensor

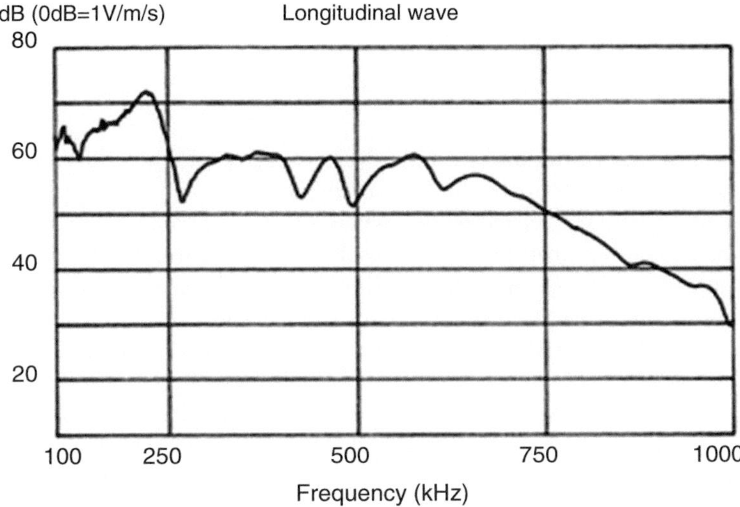

Fig. 3.4 Frequency characteristics of a 150 kHz resonance-type AE senor

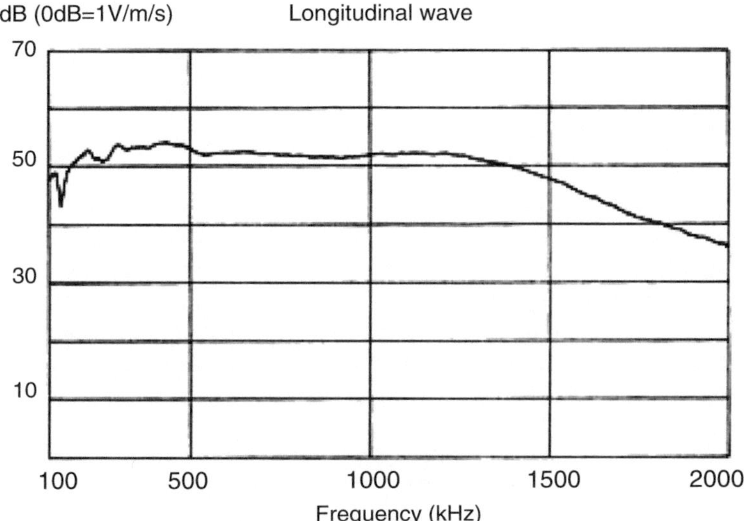

Fig. 3.5 Frequency characteristics of a broad-band AE sensor

The transfer element is fixed to the inside surface of the alumina with an adhesive. Figure 3.5 shows the sensitivity–frequency characteristics of a broad-band AE sensor with a frequency range from 100 kHz to 1.3 MHz. The unit of the vertical axis (sensitivity) is decibels assuming that 0 dB corresponds to 1 V/m/s.

3.4   Various Specifications of AE Sensors

3.4.1   Insulation (Surface for Installation)

When there is a potential difference between the ground where the AE sensor is placed and the ground where the measurement instrument is placed, there is a potential difference between the two ends of the signal cable connecting the AE sensor and measuring instrument. As a result, there is a current in the signal cable. This current acts as noise since the AE signal current is passing through the same cable. Figure 3.6 shows the principle of noise generation.

When a number of AE sensors are used at the same time, measurement will be affected by noise caused by a magnetic field because the signal cables between AE sensors and a measurement instrument form a loop circuit like a coil to generate current even though there is no potential difference. As a countermeasure for this noise, alumina, known to be an insulator, is widely used for the detection face of an AE sensor. Depending on how the AE sensor is fixed and what type of jig is used, electrical conduction between the AE sensor and testing object may short circuit even though the mounting face of the AE sensor is insulated. For this reason, it is necessary to use an insulated jig dedicated to fixing the AE sensor or to fix the case of the AE sensor with adhesive tape.

3.4.2   Waterproofness

If water enters the housing of the AE sensor, not only will the measurement be affected but also the AE sensor might be damaged. Therefore, it is recommended to use an AE sensor that is at least drip-proof and possibly even waterproof in certain environments. Even though there might be no damage due to water, there is the possibility of a short circuit due to water in the case that only the detection face is insulated. To avoid such a short circuit, it is recommended to use a fully waterproof

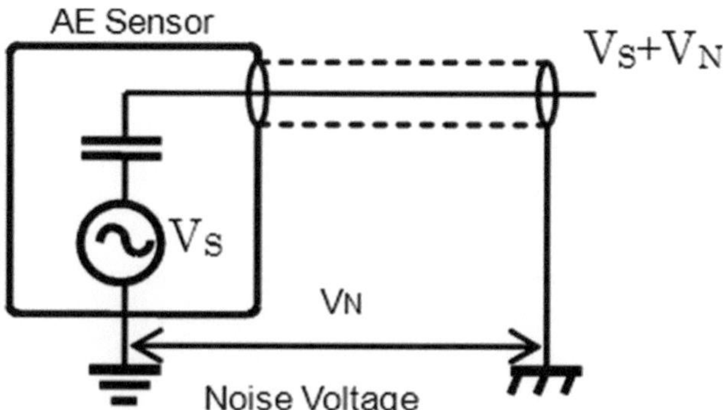

Fig. 3.6 Noise from a potential difference of the ground

and insulated AE sensor for which the whole surface of the case is insulated from the signal line. Such a waterproof AE sensor is not equipped with a signal output connector but with a directly fixed output cable.

3.4.3   Signal Output Method

An AE signal is very weak and has a magnitude of only a few millivolts at the output terminal of the AE sensor. Consequently, noise from the outside must be blocked so as to reduce interference as much as possible. For a cable connecting an AE sensor and preamplifier, a special (low-noise) coaxial cable is usually used. A metallic shield covering the outer side of the special (low-noise) coaxial cable shelters the AE signal from noise and the AE signal passes through the center core within the dielectric insulator of the coaxial cable. This signal transmission method is referred to as the single-ended (unbalanced) method. Even if the above method is used, noise might be induced at the outer conductor of a coaxial cable when the noise outside the cable is strong. As a result, the signal includes the noise at the outer conductor. To address this problem, the differential (balanced) method is employed, where two-core coaxial cables with respective metallic shields are used instead of a normal coaxial cable. Antiphase AE signals are designed to pass through their respective internal cores at the same time. The AE signals are then amplified and combined while in opposite phase. In differential (balanced) transmission, external noise is erased because it appears in both the signals of normal phase and opposite phase with equal amplitude Vn (called common mode noise). In this case, it is necessary to prepare a specified amplifier with differential input to drive the differential (balanced) AE sensor. The principle of signal transmission for differential (balanced) output is shown in Fig. 3.7.

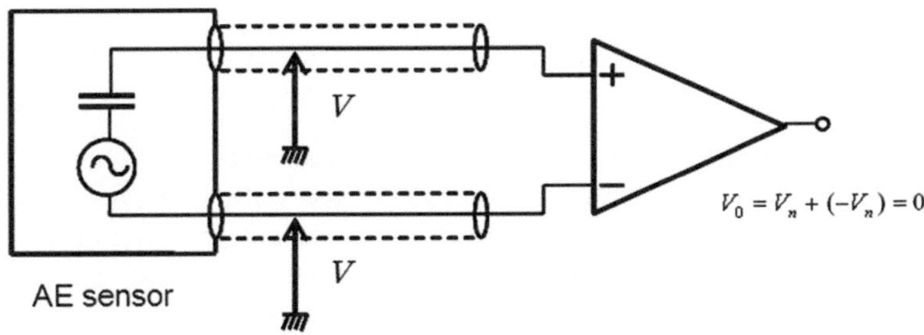

Fig. 3.7 Schematic diagram of differential/balanced output

3.4.4   Cable (Low-Noise Cable)

For use with an AE sensor, a special coaxial cable called a low-noise cable is recommended. When a signal cable is bent, twisted and/or shaken, the cable generates electrostatic noise. This is because the motion of a cable forms a local capacitor as a result of separation of the external conductor and insulator, which introduces noise at the time of discharge of an electrical charge. To prevent such noise, a special low-noise cable is widely employed, where a conductive layer (carbon) is glazed on the surface of a dielectric insulator.

3.4.5   Integrated Preamplifier Sensor

A conventional AE sensor outputs a signal generated at its transfer element directly to a preamplifier, while another type of AE sensor is available with a preamplifier installed within its case. In the latter case, the AE sensor amplifies AE signals by 20–40 dB (a factor of 10–100) as to transmit a strong signal. In other words, signals are transmitted upon completion of the damping of impedance within the AE sensor. Thus, the AE signals are noise-resistant as they are not affected by noise from the outside. Furthermore, the system is available even for a long distance between the AE sensor and a preamplifier and/or a measurement instrument, because little signal attenuation is expected. It is also possible to use a normal coaxial cable as a signal cable when the AE sensor with built-in preamplifier is used. To supply power to an AE sensor with a integrated preamplifier, a signal output cable is usually used. An AE sensor integrated with preamplifier has varying specifications such as a power supply voltage from 15 to 24 V and impedance of 50 or 75 Ω. Accordingly, it is necessary to use an appropriate measurement instrument such that the power supply meets the specification of the AE sensor.

3.5   Calibration of AE Sensors

Depending on how an AE sensor is used or the circumstances in which it is used, there may be sensitivity deterioration and/or changes to the frequency characteristics of the AE sensor. The main factor is a change in the adhesive condition between the detection face and the piezoelectric element. The adhesive condition can change as a result of repeated changes in the external load and temperature. It is thus necessary to confirm that the properties of an AE sensor.

1. NDIS2109: Methods for absolute calibration of acoustic emission transducers employing the reciprocal technique (reciprocal calibration method)

   Method for Absolute Calibration of Acoustic Emission Transducers by Reciprocal Technique published by the Japanese Society for Non-Destructive Inspection

Standards is available for the calibration of an AE sensor. The standard specifies the following procedure.

First, prepare three AE sensors. Receiving and transmitting signals for the three sensors placed on a designated block are used for calibration. The AE sensor's sensitivity to physical volume (velocity) is then obtained by calibrating the sensitivity for a longitudinal wave and that for a surface wave. Furthermore, the standard specifies the following.

As a secondary calibration method, a single calibrated AE sensor can be used as a standard to calibrate another AE sensor with/without a integrated preamplifier though number of calibrated AE sensor is only one(1) piece.

2. ISO 12713: Primary calibration method for transducers

ISO 12713: Primary calibration method for transducers regulates the calibration of sensitivity to the physical volume, as an alternative to the reciprocal calibration method.

The procedure is as follows.

First, lay a glass capillary tubing on a steel block. Then press the glass capillary tubing downward so that the surface of the steel block is pushed down gradually. When the glass capillary tubing breaks, the surface of the steel block recovers from a compression to a flat shape. At the very moment of displacement, the steel block emits a signal. Making use of the displacement of the steel block as a signal source, calibrate the sensitivity of the AE sensor by comparing with the displacement already acquired using a capacitive displacement sensor. It has been reported that there is good correlation between the results of the reciprocal calibration method and primary calibration method for the same AE sensor.

3. Method for checking sensitivity deterioration

It is difficult for users of AE sensors to work out the reciprocal calibration method or primary calibration method by themselves because both methods require a large steel block and dedicated facilities. Therefore, it is natural for users to ask manufactures about the calibration of AE sensors. There is a simple method for users to confirm whether the sensitivity of an AE sensor has changed. First, fix the AE sensor to a steel block with appropriate dimensions. Then record the output signal of the AE sensor while breaking the lead of mechanical pencil on the steel block. Data can then be compared with initial state in sensitivity. (Refer to *NDIS2110 Method for measurement of sensitivity deterioration of an AE transducer*) It is recommended that users ask manufactures to conduct an official sensitivity calibration. Furthermore, users are encouraged to carry out a preliminary check before they use an AE sensor.

Chapter 4
AE Measurement System

Masaaki Nakano and Hideyuki Nakamura

Abstract AE measurement system used for AE testing is presented in this chapter. Purpose and characteristics of preamplifier, frequency-filter and signal cable are explained. Functions of general AE measurement software are also described.

Keywords AE measurement system • Preamplifier • Frequency-filter • Signal cable

Figure 4.1 shows the schematic flow from AE wave generation to the measurement and output of the processed results. The region enclosed by a dotted line in the figure represents an AE measurement system. The AE measurement system generally consists of an AE sensor, a preamplifier that amplifies and transmits the AE sensor output to the AE signal processor, and an AE signal processor that amplifies and processes the AE signal in various ways, extracts useful information from the processed signal, and analyzes, displays and records the signal. As the AE sensor is described in Chap. 3, this chapter describes the preamplifier and other components.

4.1 Configuration of an AE Measuring Instrument

An AE measuring instrument is configured in various ways depending on the purpose of the measurement, the type and size of the target and the site conditions. AE measurement can be performed by the simplest configuration as shown in Fig. 4.2, in which a single AE sensor's output is connected to a general-purpose oscilloscope or an AC (effective value) voltmeter to observe AE waveforms or signal levels.

Mr. Nakano wrote this book while at Chiyoda Corporation. The original corresponding author was Mr. Nakano. The author was changed to Dr. Nakamura due to the decease of Mr. Nakano.

M. Nakano
Chiyoda Corporation, Yokohama, Japan

H. Nakamura (✉)
IHI Inspection & Instrumentation Co., Ltd., Yokohama, Japan
e-mail: hideyuki_nakamura@iic.ihi.co.jp

© Springer Japan 2016
The Japanese Society for Non-Destructive Inspection, *Practical Acoustic Emission Testing*, DOI 10.1007/978-4-431-55072-3_4

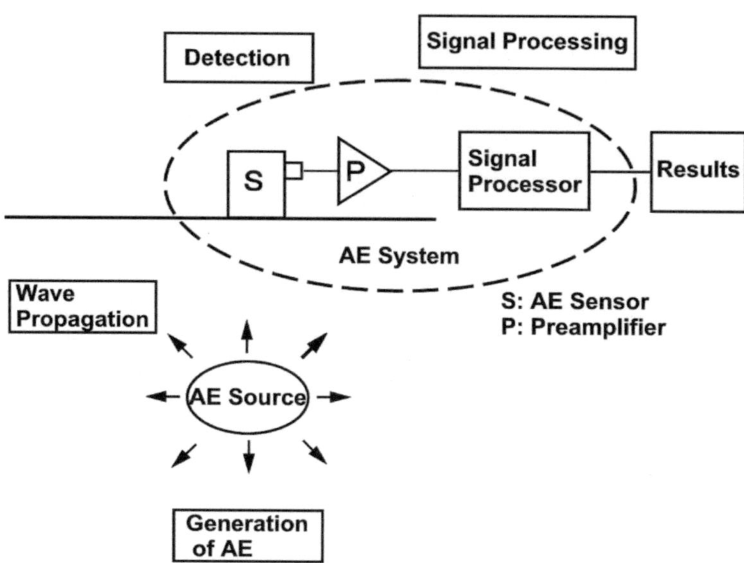

Fig. 4.1 General flow of AE measurement

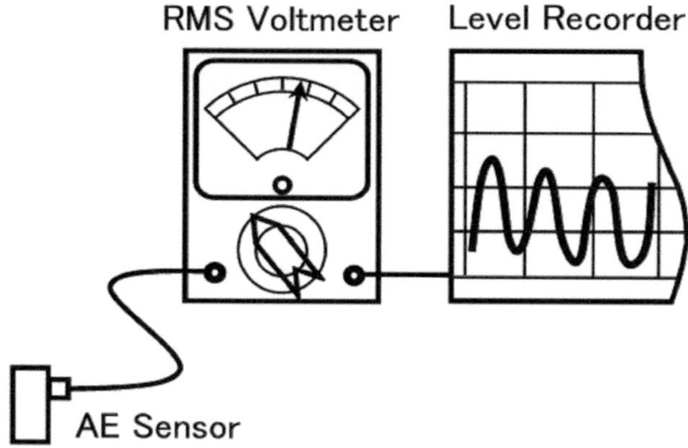

Fig. 4.2 Simple AE measurement system

On the other hand, AE testing of large structures such as plant equipment requires an AE measurement system that has many AE channels and the capability of real-time AE source location and display as shown in Fig. 4.3. This section describes the following basic configuration of the AE measurement system.

4.1.1 Basic Configuration of the AE Measurement System

The basic configuration of the general multi-channel AE measurement system is shown in Fig. 4.4.

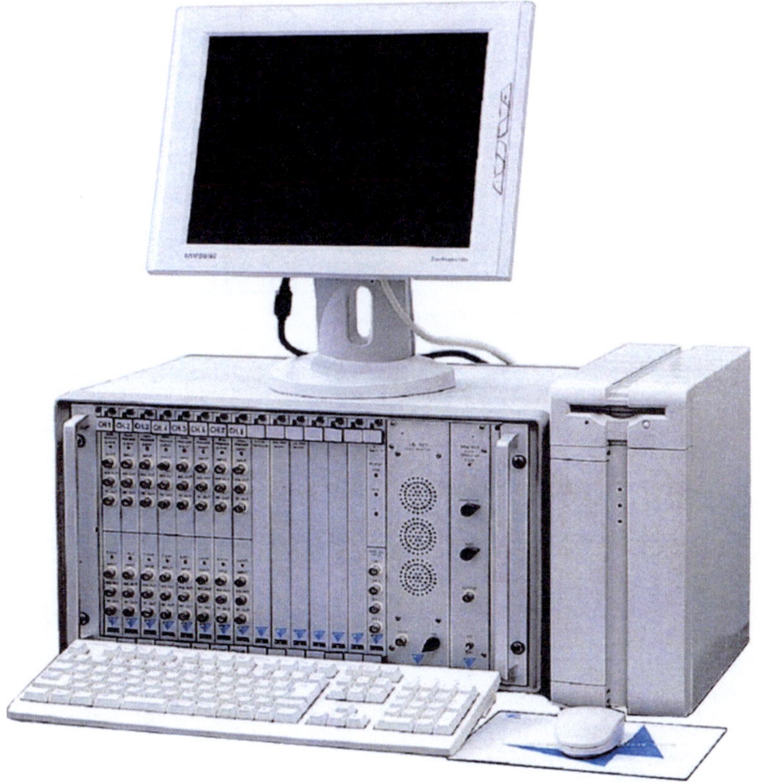

Fig. 4.3 Multi-channel AE measurement system for structural testing

The output signal from the AE sensor is amplified by the preamplifier and input into the AE signal processor. In the AE signal processor, the AE signal is filtered by a frequency filter and amplified by the main amplifier. The various AE parameters are then extracted from the signal in digital form with a combination of an A/D converter and a digital signal processor (DSP) or through an AE parameter extraction circuit and transmitted to a computer. The computer analyzes the data with appropriate software, and the results are output and displayed. The AE data are stored in the computer for future analysis.

4.1.2   Preamplifier

The preamplifier amplifies the output signal from the sensor and drives the cable to the main amplifier. The preamplifier is necessary because the amplitude of the AE sensor output signal is small and the impedance of the signal source is high, such that the signal from the AE sensor is not suitable for driving a long cable and is

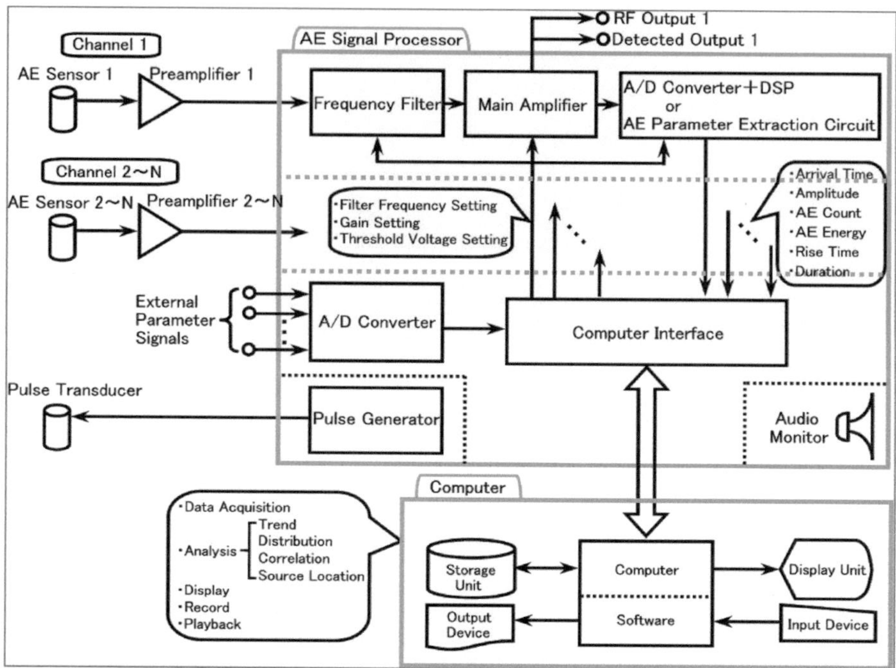

Fig. 4.4 Basic configuration of the AE measurement system

susceptible to noise. In selecting a preamplifier, it is necessary to consider issues such as the input/output types, the gain (amplification ratio), frequency characteristics, input/output impedances, power supply, shape/dimension/weight and environmental impact, depending on the intended purpose. The specification items are summarized as follows.

4.1.2.1 Input/Output Types

There are two types of signal transmission: balanced (also called as differential-type) and unbalanced (also called single-ended) transmissions, as illustrated in Fig. 4.5.

In an environment in which there is a high level of electromagnetic noise, balanced transmission improves the signal immunity to in-phase (common mode) noise. As mentioned in the previous chapter, as an AE sensor output can be of either balanced or unbalanced type, the input type of the preamplifier should be adapted correspondingly. There is a preamplifier that can switch between balanced and unbalanced inputs. The output type of the preamplifier should also be adapted to that of the main amplifier; however, because the signal level at the output of the preamplifier is considered to become sufficiently large, balanced transmission is

Fig. 4.5 Balanced (*upper*) and unbalanced (*lower*) transmissions

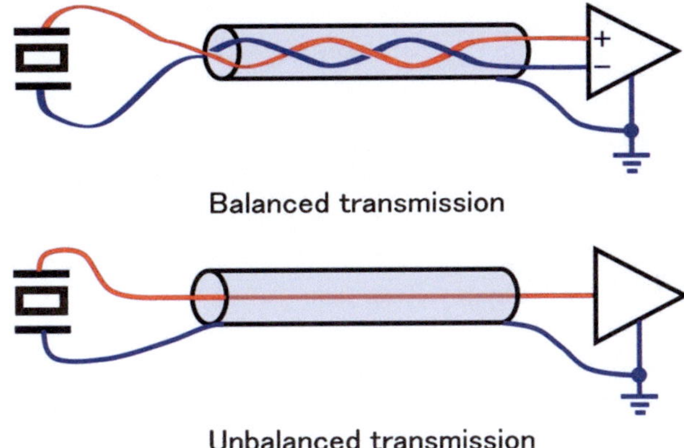

used in most cases and the preamplifier is connected to the main amplifier by a coaxial cable.

4.1.2.2 Gain (Amplification Ratio)

The output signal of the AE sensor is generally at a low level of the order of 10 μV to several millivolts. The preamplifier first amplifies the signal to facilitate subsequent processing. The optimal value for the gain of the preamplifier depends on the purpose of the AE measurement. Most preamplifiers usually have a fixed gain between 20 dB (an increase of a factor of 10) and 40 dB (a factor of 100), but there are preamplifiers that have switch-selectable gains.

Description of Term **(dB (Decibel))**
The dB (decibel) is a unit that represents the ratio of two values, and is often used in electronics and acoustics to describe a gain or attenuation. A voltage ratio is expressed as

$$A(\mathrm{dB}) = 20 \log_{10}\left(\frac{V_1}{V_0}\right),$$

where A is the voltage ratio in dB representation, $\log_{10}$ is the common logarithm, and V_1 and V_0 are the voltage to be compared and the reference voltage, respectively.

In AE measurements, a value in dB is sometimes used to express an absolute voltage relative to 1 μV. It is then denoted $\mathrm{dB_{AE}}$:

(continued)

$$B(\mathrm{dB_{AE}}) = 20\log_{10}\left(\frac{V}{1\mu V}\right),$$

where V is the AE signal amplitude at the AE sensor output (before amplification).

4.1.2.3 Frequency Response

The frequency response of the preamplifier should suit the purpose of the AE measurement. A frequency band of several 100 kHz or higher may be important in the AE measurement of some materials, while an audible frequency range of several kilohertz may be used for other materials. The preamplifier to be used with an AE sensor should have a frequency range that is fully compatible with that of the sensor. Most general-purpose preamplifiers that cover a frequency range of several kilohertz to several megahertz can be used, unless the range is otherwise specified for a special purpose. It should be noted that a broader range of frequencies leading to a higher level of noise is disadvantageous in AE measurements. There is an amplifier with a built-in band pass filter that can limit the frequency range. Figure 4.6 shows an example of the frequency response of a preamplifier.

4.1.2.4 Noise

Noise is always a problem in AE measurements. Ideally, noise should not be generated inside an amplifier; however, in practice, noise, including thermal noise from an electronic circuit, is unavoidable.

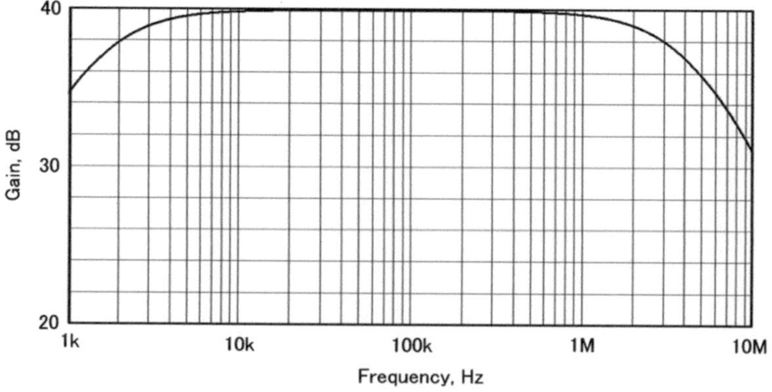

Fig. 4.6 A frequency response of a preamplifier

The preamplifier is installed at the first stage of the AE signal amplification system and the noise generated in the preamplifier is amplified thereafter, and this affects the overall signal-to-noise (SN) ratio of the AE measurement. Consequently, a lower level of noise is always better. Depending on the frequency range, the amplitude of noise is generally several microvolts root-mean-square (rms). It should be noted that a maximum amplitude of the noise (peak value) is several times larger than the corresponding rms value.

Description of Term (**SN Ratio**)
The SN ratio, or S/N, is the ratio of signal to noise and is usually expressed in dB. Large SN ratios are always preferable.

Description of Term (**RMS Voltage**)
The term rms is an abbreviation for the root-mean-square and is also called the effective value. An effective value *Vrms* in a time width T for a voltage ($V(t)$) that varies with time is expressed by

$$Vrms = \sqrt{\frac{1}{T}\int_0^T V(t)^2 dt}.$$

4.1.2.5 Input/Output Impedance

As the AE sensor is connected to the input of the preamplifier, higher input impedance is preferable. In particular, the equivalent input capacity should not be too large. The output impedance of the preamplifier should be much smaller than the input impedance of the main amplifier. The length of the cable that connects the output of the preamplifier to the input of the main amplifier (a coaxial cable in most cases) may exceed 100 m in some cases. For this reason, the input terminal of the main amplifier should be terminated (matched) by a resistance with the same characteristic impedance as that of the coaxial cable (often 50–100 Ω). Consequently, it is recommended that the output impedance of the preamplifier should be a few ohms or less. When the output impedance of the preamplifier and the input impedance of the main amplifier are matched with the characteristic impedance of the coaxial cable, it should be noted that the input voltage of the main amplifier becomes half the open circuit output voltage of the preamplifier.

Description of Term (**Impedance**)
Impedance is the ratio of voltage to current for alternating current. More generally, impedance $\dot{Z}$ is

(continued)

$$\dot{Z} = \dot{V}/\dot{I}$$

where $\dot{V}$ and $\dot{I}$ are the voltage and current in consideration of phase, respectively.

Description of Term (**Characteristic Impedance**)
Coaxial cables and coaxial connectors for high-frequency use are designed to have characteristic impedances. For many coaxial cables, this characteristic impedance is 50–100 Ω.

Description of Term (**Impedance Matching**)
Impedance matching is the equalizing of impedances on the signal-transmitting side and signal-receiving side. Impedance matching is necessary for maximum efficiency in power transmission. A coaxial cable and coaxial connector with the same (matching) characteristic impedance should always be used. Impedance mismatching causes reflection of the signal and distortion of the waveform.

Description of Term (**ASL**)
The ASL is an average signal level and a time average value of the detected (rectified) AE signal:

$$\text{ASL} = \frac{1}{T}\int_0^T |V(t)|dt.$$

4.1.2.6 Power Supply

A power supply is necessary to operate a preamplifier. Because the maximum output voltage is limited by the supply voltage of the preamplifier, a large supply voltage is preferable for a wide dynamic range; however, it should not be too high for reasons of safety and handling. Generally, the supply power is around 15–30 V. Some preamplifiers require bipolar power (e.g., ±15 V). There are two methods for supplying power to the preamplifier: the use of a dedicated cable and the superposition of a supply power on a signal cable. The latter case is more convenient as only one cable is needed.

4.1.2.7 Dimensions/Shape/Weight and Environmental Resistance

It is necessary to install the preamplifier close to the AE sensor; i.e., on the measurement object or as close as possible to the object. For this reason, the dimensions, shape and weight of the preamplifier are important. The preamplifier should be small and light. There is an integrated type of AE sensor with a preamplifier, and also a preamplifier with a built-in AE sensor. Compatibility with the environmental conditions is also an important issue. A preamplifier should be resistant to various environmental conditions such as the presence of water and oil and extreme temperature.

4.1.3 Main Amplifier

The main amplifier, contained in the chassis of the AE measurement system, receives signals from a preamplifier, amplifies the signals further, selects and passes signals in necessary frequency bands through a filter, and cuts off signals in the unnecessary frequency bands. The main amplifier normally has output terminals for waveform output (high-frequency) signals and detected signals that are connected to peripheral devices. The gain at the amplifier is usually some 10–40 dB. For the filter, a bandpass filter that cuts off low-frequency and high-frequency bands is often used. The cut off frequencies are set depending on the purpose of the AE measurement and in consideration of the AE sensor's frequency characteristics. In the case of metallic materials for AE measurement, the filter is often set to pass a signal with a frequency of several 100 kHz, but in the case of concrete and rock, a range of some 10–100 kHz is sometimes set as the pass band. In addition to the cut-off frequency, the characteristics of the filter include the attenuation slope, phase characteristics, and transient characteristics, but it is only pointed out here that a filter with a steep attenuation slope (steep cut off characteristic) tends to impair the phase and transient characteristics. The external outputs of both the high-frequency and detected signals should have sufficiently low impedances. A detected signal is often output through a low-pass filter ranging from several hertz to the order of 10 Hz.

In general, an up-to-date digital measurement system directly digitalizes signals input from a preamplifier without a main amplifier.

4.1.4 Signal Processor (AE Parameter Extraction Circuit)

The AE signal processor (AE parameter extraction circuit) extracts various AE parameters described in Sect. 2.5.4 from the amplified and filtered AE signals.

The extraction is accomplished by digital processing with an A/D converter and a DSP, or by an individual signal processing circuit. For burst AE signals, as schematically shown in Fig. 4.7, an AE hit is recognized when the AE signal exceeds a set voltage threshold. Each AE parameter is extracted for one AE hit and transmitted to the computer. Each AE hit data contains the arrival time (time stamp), and a multi-channel AE measurement system locates the AE source in one, two or three dimensions according to the arrival time difference between channels.

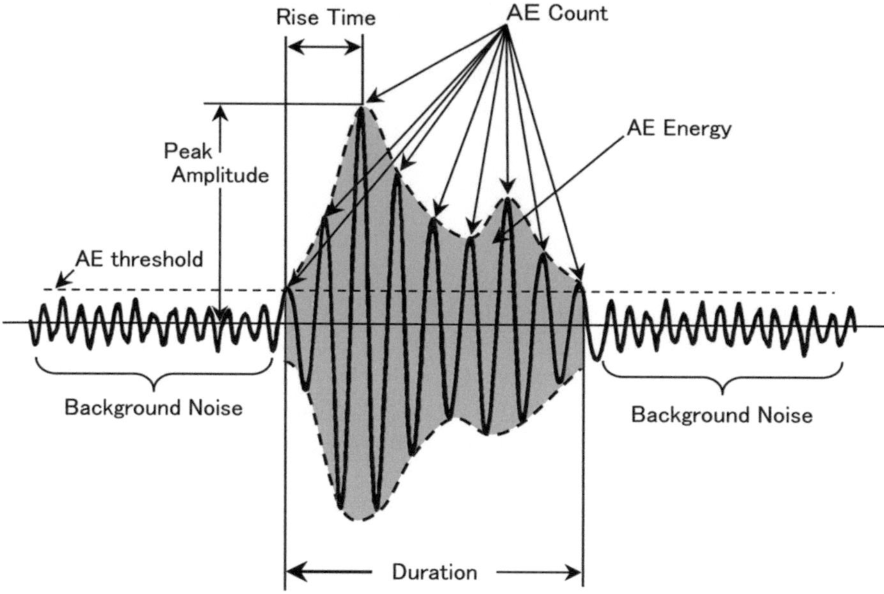

Fig. 4.7 AE threshold and AE parameters

On the other hand, in the case of continuous AE signals, such AE parameters are not defined and only the signal level (RMS voltage or ASL) is measured.

4.1.5 Output/Display Device (Computer for Measurements)

AE data output from the AE signal processor is transmitted to a computer for measurement through the computer interface. The computer executes various analyses of the data using AE measurement software and outputs/displays the results. The AE measurement software has the functions listed in Table 4.1. In most cases, the software is able to execute all or some of these functions in real time. Figure 4.8 is a screen shot of the AE measurement system.

Because the functions and performance capabilities required of the AE measurement system vary widely, it is necessary to choose an AE measurement system that is appropriate for its purpose. If the processing speed is insufficient, the system may lose important AE signals or discontinue operation in the case of a very high occurrence rate of AE.

Specialized software for many AE measurement systems can be executed on general-purpose operating systems such as Microsoft Windows with a user interface similar to that used in general application software. However, it is recommended that other applications should not be run simultaneously during the execution of the AE measurement software for real-time processing. In particular,

Table 4.1 Functions of typical AE measurement software

Function		Description
Data acquisition		Acquires measured AE data from AE signal processor to computer
Data analysis	Trend	Traces time history of AE parameters
	Distribution	Analyzes distributions of AE parameters
	Source location	Performs source location calculations in linear, planar or 3D modes
Display		Presents analyzed results on display as graphs and tables
Record		Records AE data in storage unit
Playback		Reproduces stored AE data

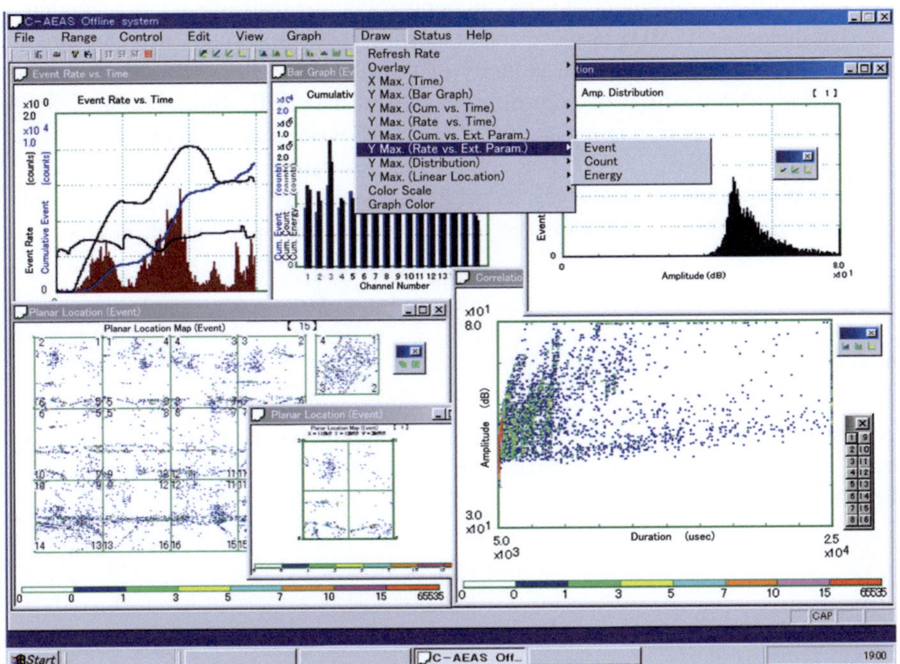

Fig. 4.8 A screen shot of the AE measurement system

programs that consume much of the central processing unit resource should be avoided.

Most modern computers are considered to be capable of operating AE measurement systems. In the case of long-term continuous AE measurement, however, means for the backup of proper data should be prepared with consideration given to the reliability of the computers involved. In some cases, it is necessary to implement the redundancy of a hard disk drive and also to prepare measures against the loss of the electric power supply to AE measurement systems using an uninterruptible power supply.

4.2   Peripheral Equipment

In AE measurements, peripheral equipment as listed in Table 4.2 is sometimes used to make up for missing functions of the AE measurement systems. Some AE measurement systems have the functions of this peripheral equipment.

4.2.1   Oscilloscope

An oscilloscope is a general-purpose measurement system to observe waveforms. For waveform measurement, it is recommended that the instrument has a frequency band of the order 10 MHz or more and at least two channels in order to measure the arrival time difference and the velocity of the AE wave. Recently, more oscilloscopes have the function of recording waveform data with digital storage. Some have the function of frequency analysis (i.e., fast Fourier transformation). There are instruments for measuring AE that can present waveforms on a display of the AE measurement computer; however, an oscilloscope that operates independently of a computer is useful in confirming the operation of AE sensors and amplifiers.

4.2.2   Spectrum Analyzer

A spectrum analyzer analyzes the frequency components of AE waves and should have a frequency band of ~1 MHz. An oscilloscope with this function can also be used.

4.2.3   Pulse Generator (Pulsar)

A signal source that artificially generates AE waves is necessary in AE measurements and in the installation of AE sensors. When a pulse signal is applied to a piezoelectric element, a mechanical strain is generated and this can be used as a simulated AE source. A device that generates electric pulse signals for this simulated source is a pulse generator. A simulated AE source produced by the pulse generator makes it possible to continuously generate a number of simulated AE signals with constant amplitude and to set and control the repetition rate and amplitude in wide ranges.

Table 4.2 Peripheral equipment for AE measurement

Equipment	Function
Oscilloscope	Waveform monitoring and recording
Spectrum analyzer	Frequency spectrum analysis
Pulse generator	Generation of test pulse
Audio monitor	Sound monitoring with loudspeaker
External parameter instrument	Measurement of load, strain, temperature and pressure, etc.
Environmental instrument	Measurement of environmental conditions such as velocity and direction of wind, etc.

4.2.4   Audio Monitor

Using an audio monitor, the AE signal, which normally has a frequency within the ultrasonic wave region, is converted to a signal with an audible frequency and output from speakers. Judging a sound from the audio monitor by ear in an intuitive manner is an effective method of checking what AE signal is detected. An experienced operator can detect and distinguish electrical noise immediately with an audio monitor.

4.2.5   Instrument for Measurements of External Parameters and Environmental Conditions

In the measurement of AE signals, data such as load, strain, temperature, and pressure data are simultaneously measured and recorded. These data, called external parameters, are measured and recorded through an A/D converter in the AE measurement system. Since specifications for the output signals of a sensor for each external parameter widely vary, it is necessary to match these signals with the specifications of the A/D converter. A signal converter for this purpose is an external-parameter measurement instrument. For instance, as in the cases of a load cell amplifier for loads and a strain amplifier for strains, appropriate converters for different sensors are used.

In the case of AE measurement at a plant site, it is sometimes necessary to measure and record environmental conditions such as the wind direction, wind speed and atmospheric temperature. Output signals of the sensors are also converted to proper signals and then input into an A/D converter to determine external parameters.

4.3   Connection Cable and Connector

A coaxial cable for high-frequency use is mainly used in AE measurement. The cable between the AE sensor and the preamplifier is generally a low-noise cable with small diameter. A general-purpose coaxial cable is used for the preamplifier-to-main amplifier connection and the connection between the main amplifier and peripheral devices.

4.3.1   Cable Between an AE Sensor and Preamplifier

An AE sensor can be considered as an electrostatic capacitance that generates charge. Because a larger equivalent electrostatic capacity of the cable reduces a preamplifier's input voltage, it is necessary to use the shortest possible cable. A 1-m cable sometimes reduces the signal voltage by half or more. Consequently, it is desirable that the length of the cable should be several tens of centimeters or less.

Because bending and vibrations along a cable are sources of noise, it is necessary to use a coaxial cable with a structure that suppresses the generation of such noise (i.e., a low-noise cable). In general, a microdot type or SMA (Sub Miniature version A) connector is used on the AE sensor side, while a BNC (Bayonet Neill Concelman) connector is used on the preamplifier side. Figure 4.9 shows an example of a cable between an AE sensor and preamplifier. Figure 4.9a shows a cable for an unbalanced AE sensor. The cable and connector shown in Fig. 4.9b are used for a balanced (differential) AE sensor.

4.3.2   Connection Cable Between a Preamplifier and Main Amplifier

In most cases, a coaxial cable with characteristic impedance of 50 or 75 Ω is used for a connection cable between the output of the preamplifier and the input of the main amplifier. A length of this cable may sometimes exceed 100 m.

For a long cable, attenuation by the cable should be considered. A thicker cable reduces attenuation, but increases the time and work involved in preparation, particularly in multi-channel measurement. In some cases, a multi-core coaxial cable that contains many (e.g., 10) coaxial cables in one bundle is used. Figure 4.10 shows an example of a 10-core coaxial cable. For this cable, a BNC connector is usually employed.

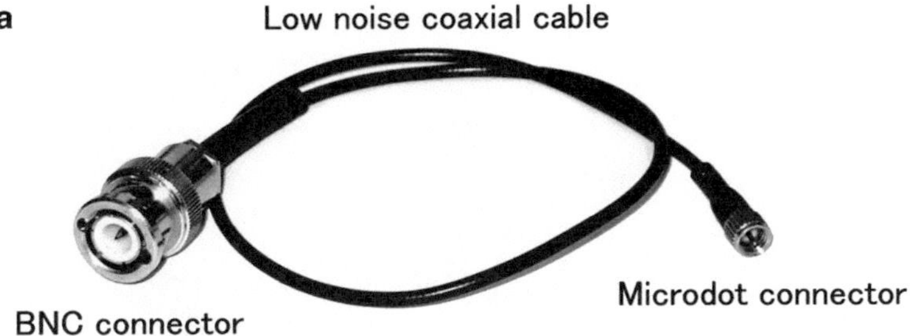

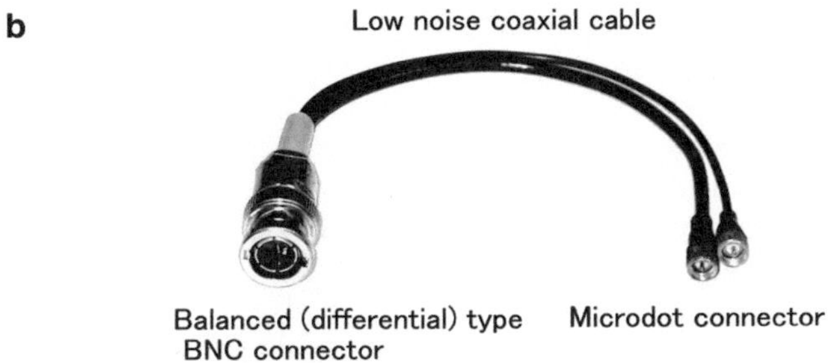

Fig. 4.9 (**a**) Cable for connecting an AE sensor and preamplifier [unbalanced type]. (**b**) Cable for connecting an AE sensor and preamplifier [unbalanced (differential) type]

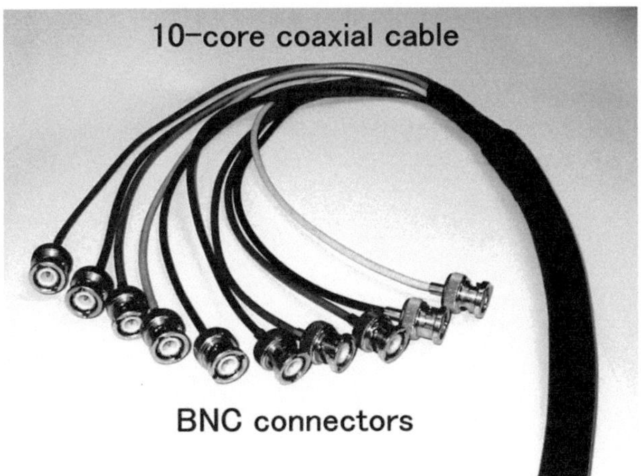

Fig. 4.10 Ten-core coaxial cable

4.3.3 *Other Connection Cable*

Output signals from the AE measurement system are supplied to various peripheral devices via connection cables. For this connection cable, a coaxial cable with BNC connectors at both ends, as shown in Fig. 4.11, is generally used.

Fig. 4.11 Coaxial cable with BNC connectors at both ends

External parameter signals are normally low-frequency signals. For this reason, non-coaxial cables such as a twisted pair line are often used as connection cables. Connectors also depend on the devices to be connected.

Connection cables (connectors) for computers are standardized according to different purposes; e.g., EIA-232 (serial communication) and IEEE1284 (parallel port) cables. Consequently, connection cables that are compatible with their intended purposes should be used.

Chapter 5
Practical AE Testing, Data Recording and Analysis

Tomoki Shiotani, Yoshihiro Mizutani, Hideyuki Nakamura, and Shigenori Yuyama

Abstract In this chapter, the practical procedure of AE testing including data recording and analysis are presented. As for the test preparation, setup of the system and sensitivity checkup ways are explained. Load application methods for AE testing are discussed, followed by some typical data-display methods.

Keywords Practical procedure of AT • Data acquisition • Loading procedure • Test preparation • Data display • Evaluation report

5.1 Preparation for AE Testing and Sensitivity Check of the AE Measurement Instrument

Hideyuki Nakamura

This section describes the setup of an AE measurement instrument, the installation of AE sensors, and a sensitivity check of the instrument. The sensitivity check in this section is not a calibration of the absolute sensitivity of the instrument, but rather, a simple sensitivity check that is part of the preparation for AE testing.

T. Shiotani (✉)
Kyoto University, Kyoto, Japan
e-mail: shiotani.tomoki.2v@kyoto-u.ac.jp

Y. Mizutani
Tokyo Institute of Technology, Tokyo, Japan
e-mail: ymizutan@mes.titech.ac.jp

H. Nakamura
IHI Inspection & Instrumentation Co., Ltd., Yokohama, Japan
e-mail: hideyuki_nakamura@iic.ihi.co.jp

S. Yuyama
Nippon Physical Acoustics, Ltd., Japan, Tokyo
e-mail: yuyama@pacjapan.com

© Springer Japan 2016
The Japanese Society for Non-Destructive Inspection, *Practical Acoustic Emission Testing*, DOI 10.1007/978-4-431-55072-3_5

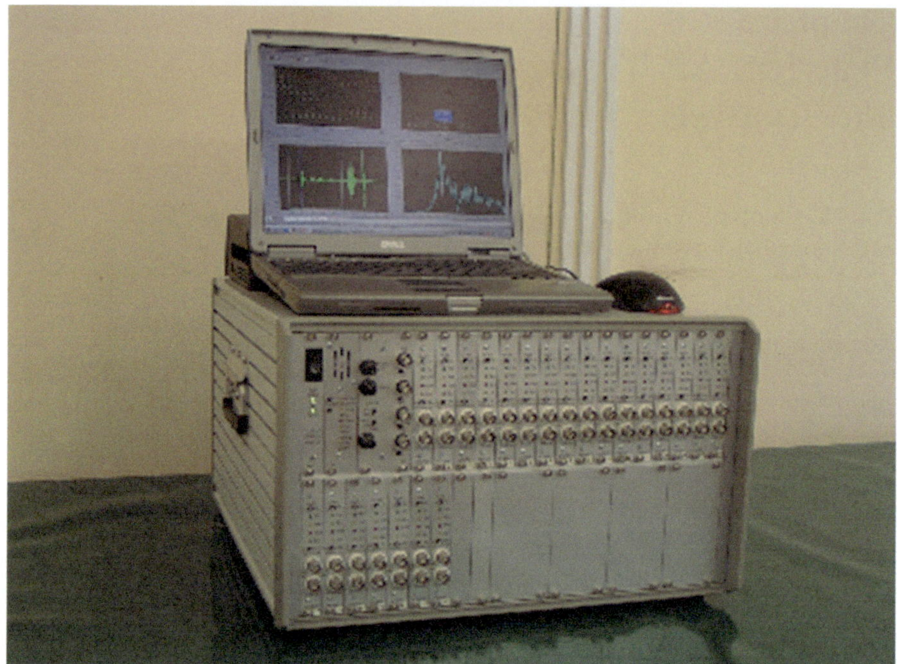

Fig. 5.1 Appearance of an AE measuring instrument

5.1.1 Setup of the Instrument

Preparation for AE testing requires not only concrete tasks such as the installation and mounting of the AE instrument and peripheral devices, but also the setting of intangible conditions such as the measurement conditions indicated in the procedures. Procedures for the setup of the instrument, along with points to be noted for different tasks, are given below.

5.1.1.1 Installation of the Instrument

An AE measuring instrument, as shown in Fig. 5.1, is installed near a test object.

Before installing the AE instrument, the effects of vibrations and electric noise must be considered. Since any vibration can be causes to fail the AE instrument or to be an electric noise, the instrument must either be installed in a place without vibrations or vibration-proofing measures must be taken.

In some cases, electric noise generated by external devices can enter the measuring instrument via a power line, and noise propagates through the air as a electromagnetic wave. Consequently, attention should be paid to peripheral devices such as welders, and rotating machines with motors and solenoid valves.

Fig. 5.2 Device-to-device cable connection

5.1.1.2 Connection of the Instrument

As shown in Fig. 5.2, cables connect the main unit of the AE instrument and devices such as an analysis computer and monitor. Power cables are connected to the individual devices.

When any electric noise enters the instrument via a power cable, the noise must be reduced using a line filter or a noise cutoff transformer. After the AE instrument has been connected, it must be operated without being connected to the sensors to check for noise intrusion.

5.1.1.3 Connection of Signal Cables

Signal cables are connected to the AE measurement instrument (refer to Fig. 5.3).

When signal cables swing or rub against each other, their contact sometimes creates acoustic noise, and in some cases generate electric noise. Furthermore, when a cable is pulled, there can be poor contact or even disconnection, and electric noise can be generated. In the case of the connection of a signal cable, it is thus important to keep the area surrounding the connector free of any possible source of applied force, and to also hold the cable at key points with adhesive tape or a cable mount.

Note that signal cables should be arranged so as not to block nearby foot-traffic. They also should not be placed beside devices that generate electromagnetic waves.

5.1.1.4 Setting of Measurement Conditions

The setting of AE measurement conditions is a critical task that affects the results of the measurement. Consequently, a Level 1 engineer must set the measurement

Fig. 5.3 Connection of
signal cables

Fig. 5.4 Setting of
measurement conditions

conditions in accordance with NDT instructions prepared by engineers of Level 2 or
higher (refer to Fig. 5.4).

When setting the conditions, it is recommended to record the set values as the
process can be repeated so that analysis proceeds readily.

The main measurement conditions to be established before measuring signals
are given below.

– Effective channels

The channels used for AE monitoring must be selected.

– AE threshold

A threshold for the identification of AE signals must be set. This threshold is set
to a level exceeding the amplitude of the background noise, and to a level at which

the necessary signals can be obtained. In general AE devices, the signal strength is converted to dB (decibels); consequently, the threshold is also expressed in dB units.

– Frequency filter

Different filters are employed according to their measurement applications: a low-pass filter that allows only frequencies lower than a set value to pass, a high-pass filter that only allows frequencies higher than a set value to pass, and a bandpass filter that passes only specific frequency bands. Some devices have no setting functions. In some cases, these filters are incorporated into a preamplifier and other devices, so it is important to confirm the specifications of the devices to be used before actual measurements.

– Hit discrimination time

The time for hit discrimination, in which different AE equipment manufacturers refer to differently, refers to a time range in which the instrument determines whether or not an input AE wave has been acquired as a hit.

When the input signal crosses the threshold, the instrument starts the time for hit discrimination. If the signal again crosses the threshold within this time, the instrument recognizes that the same AE wave continues and restarts the time for hit identification. The instrument recognizes that a hit has been completed when no signals crossing the threshold are input after the lapse of the time for hit discrimination.

The time is determined according to the size and material of the measurement object, and an assumed frequency of AE occurrence.

– Peak recognition time

The peak recognition time refers to a time at which the instrument identifies the peak amplitude in one hit. With the input of a signal crossing the threshold into the AE measuring instrument, the instrument starts a time period for peak recognition. When an amplitude larger than the previous peak is input into the instrument within that time, the instrument restarts the time for peak identification. In a case that no signals exceeding the previous peak amplitude are input into the instrument, a peak amplitude and rise time in the hit are determined. Some AE instruments automatically calculate the time without setting the peak recognition time.

– Dead time

Dead time refers to the time between after the completion of the time for hit discrimination, and the time when the data capture restarts. This dead time is set to eliminate both of reflected waves and delayed waveforms to detect only effective AE waveforms. The term, used to refer to the dead time varies from one manufacturer to another.

– Gain

The gains, namely amplification rates of a main amplifier and preamplifier that are actually connected are input as a set value.

– Selection of a floating threshold

General AE devices have the function to automatically change a threshold in correspondence with variations in background noise. They equips the switch whether this function is enabled or not. When it is enabled, it is necessary to set the condition of the change. Normal AE measurements are conducted with a fixed threshold.

– Recording waveform data

In the case that waveforms are recorded it is necessary to set pretriggers, a sampling speed, a waveform length, and other parameters as described in the following. Some AE devices have an additional function to determine if the signal are recorded or not based on AE parameters.

– Setting of pretriggers

Waveforms are recorded from a starting point when AE waves exceeding a threshold to a time before the starting point set as the pretrigger.

– Setting of sampling frequency for waveform data

Waveform data is a collection of amplitude points with constant time interval. For this setting, the time internal of the point (sampling interval, inverse of sampling frequency) is decided.

– Selection of waveform length

The length of a waveform that is recorded as one waveform is selected. In general AE devices, the length is set by the number of samples per waveform.

– Selection of AE parameters

In some AE devices, one must first select the AE parameters to be acquired. In other AE devices, acquisition conditions can be changed according to the individually set AE parameters conditions.

Typical AE parameters include the amplitude, count, rise time, duration, and energy.

– Setting for input of external parameters

General AE devices have a terminal for the input of external measurement data such as stress and strain, and can simultaneously sample the data and AE parameters. The input of external parameters requires the setting of a voltage range and a coefficient.

5.1.1.5 Setup of the Monitor Display

A diagram for the monitoring of AE data in the measurement is drawn.

– Display of AE parameters

The data name, unit, graph type, and display range for each axis in the diagram are set. When displaying AE parameters, general AE instruments have a function to set various types of filters.

– Display of results of the AE source location

When implementing AE source location, the shape of the target object, the number of sensors to be used, the coordinates of the sensor(s), the size of the object, its wave velocity, and other properties are set on a screen for location conditions. To display the results of source location on the monitor, one must give settings to display a graph, including the results of the above calculation.

5.1.2 Installation of an AE Sensor

In AE testing, AE waves that propagate from a measurement object to an AE sensor placed on the object are converted to electric signals. For this reason, elastic waves must be able to be freely transmitted in the space between the object and the sensor, and must always be reproduced without variations during measurement. Detailed procedures for the installation of the sensor, and points to be noted, are given below.

5.1.2.1 Selection of Position for Sensor Installation

The position for sensor installation is selected with consideration of the AE location, propagation path and mode, the condition of the installed surface, and the expected attenuation of the AE waves. In particular, in the case of using several sensors, the positions where these sensors are installed are selected so as to obtain sufficient precision of source location, with consideration of the arrangement and intervals of all sensors that will be placed (refer to Fig. 5.5).

When the generation of noise is expected, one can reduce the noise by installing the sensor away from the source of the noise. One can also identify noise by installing guard sensors on the propagation route of the noise, as well as by taking other countermeasures. During AE testing conducted as a fatigue or tensile test, the measurement object is sometimes deformed, leading to detach sensors from the surface of the object, or causing a gap at the contact surface between the sensor and object. Both of these eventualities contribute to the generation of noise and a reduction in the sensor's sensitivity. In these cases, it is important to install a sensor

Fig. 5.5 Selection of position for sensor installation

in a position on the object where there is little deformation. In AE testing of large structures such as oil tanks, before deciding upon the positions where sensors will be installed, it must be confirmed that there are no impediments such as marginal welding or piping at those positions.

5.1.2.2 Pretreatment of the Surface Where a Sensor Will Be Installed

Before the installation of a sensor, the condition of the surface where it will be installed must be examined, to determine if the surface could interfere with the detection of AE waves. If this is the case, the surface must be pretreated as needed, according to the detailed procedures given below.

– Projections and depressions on a coated surface

Projections and depressions on a coated surface reduce a sensor's contact area resulting in less sensitivity. They must be eliminated using sandpaper or a scraper to achieve a flatness of the coated surface (refer to Fig. 5.6).

– Surface contaminated with rust or oil

When rust or oil adheres to the surface, the sensor's adhesion to the surface is compromised, resulting in a reduction in the sensitivity of the sensor. Rust must be eliminated with sandpaper; an oily surface should be cleaned with an appropriate cleaning solvent, and wiped clean with a disposable material.

For reference, when the sensor is attached with an adhesive, the surface must be completely smooth without any oily agent.

Fig. 5.6 Pretreatment of surface for sensor installation

Fig. 5.7 A sensor holder

5.1.2.3 Method for the Mechanical Installation of a Sensor

There are two general methods for the installation of AE sensors. One is that a sensor is mechanically attached to a measurement object. Another is that the sensor is affixed to the object by an adhesive. There are several methods of mechanical installation. One simple method is to attach the sensor to the object by adhesive tape, a rubber band, or a C-shaped clamp. Another is to use a special sensor holder that equips a magnet and spring, as shown in Fig. 5.7. A third method is to use a fixing jig fabricated to a measurement object. For the method that employs mechanical pressure, a jig must be selected whose resonance or movement does not become a source of noise. In this method, the pressure applied to the installation surface changes the sensor's sensitivity, so that it is important to keep the pressure on all of the sensors constant. For the mechanical installation of a sensor, any gap between the sensor and measurement object must be filled with a coupling medium. General coupling media include machine oil, glycerin, water and grease. Any coupling medium that does not deteriorate or run down during measurement is

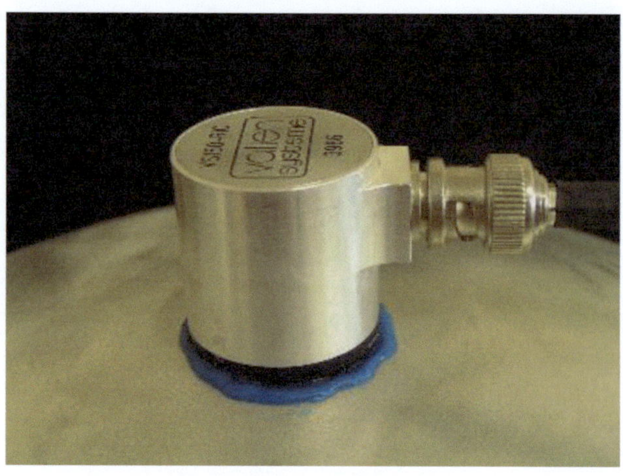

Fig. 5.8 Sensor installation with adhesives

appropriate, with consideration of the test period, surroundings, temperature of the object, and other factors. In general AE tests, such vacuum silicon grease or as a coupling medium having relatively excellent durability and high viscosity, is used.

5.1.2.4   Method for Attaching a Sensor with Adhesives

A method for attaching a sensor with an adhesive, as shown in Fig. 5.8, is used in the case that a magnet-type sensor holder cannot be used because the measurement object consists of non-magnetic substances such as resin and nonferrous metal, and fixing jigs cannot be used owing to structural considerations. The sensor could be damaged when being removed, and this must be taken into account when selecting an adhesive and removing a sensor. An adhesive attaches a sensor and also serves as a coupling medium. Consequently, a selected adhesive must not deteriorate or form gaps during measurement, and the possible removal of the sensor must be taken into account, as discussed above.

When installing the sensor, an instant adhesive that allows for easy bonding and the application of a remover is generally used. An adhesive with high bonding strength is sometimes required for tests involving vibration and strain, in this case sensor-housing materials should be sufficiently rigid to withstand any stress when it is removed, such as a sensor with a metal housing. The sensor must be removed by a remover so that the sensor and the measurement object are not damaged. When a sensor is installed on a resin or coated surface, the surface may deteriorate or deform. Consequently, it is necessary to make a preliminary test and check the effect of a given adhesive and remover on a bonded surface. Before using a given adhesive, it is also important to check that it has the same acoustic properties as a normal coupling medium.

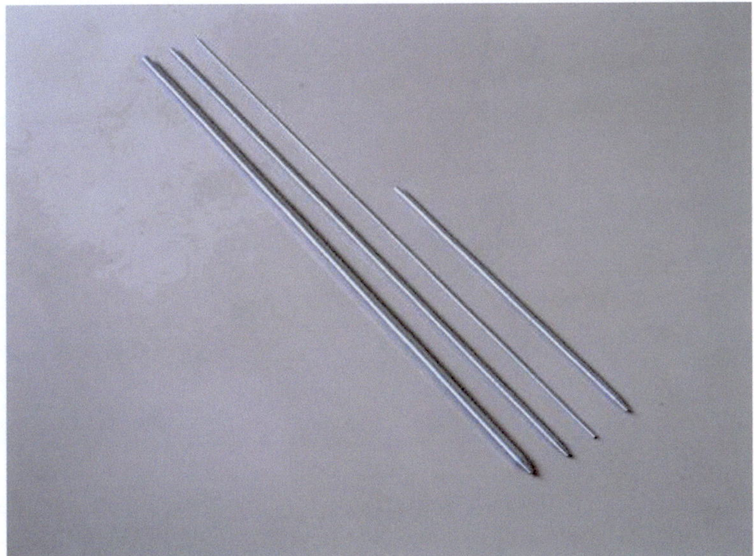

Fig. 5.9 Various types of waveguides

5.1.2.5 Waveguide

When it is difficult to place a sensor directly on the surface of a measurement object owing to structural or environmental factors, a rod-like waveguide is used to guide AE waves to a location away from the surface (Fig. 5.9). The waveguide can be permanently welded to the surface, or temporarily fixed in the same way as a sensor. When an AE signal is detected via a waveguide, certain shapes and materials of the waveguide attenuate the propagating AE wave or convert its mode. Therefore, before using the waveguide, one should conduct a preliminary test to understand its propagation characteristics.

When source location is conducted, it is necessary to consider about propagation time through the waveguide and need to adjust AE sensor locations used for the calculation. If the waveguide comes into contact with other structures, an AE wave may leak from the contact point. Consequently, the waveguide must be carefully installed so that it makes no contact with other structures. When the waveguide is inserted into objects such as thermal insulating material with which it maintains steady contact, it is important to confirm the effect of this contact in advance.

5.1.2.6 Connection of a Preamplifier

When a preamplifier such as that shown in Fig. 5.10 is used, a shorter cable between the sensor and preamplifier is conducive to the reduction of electric noise. However, a preamplifier installed on the side of a testing machine or rotary equipment may

Fig. 5.10 A preamplifier

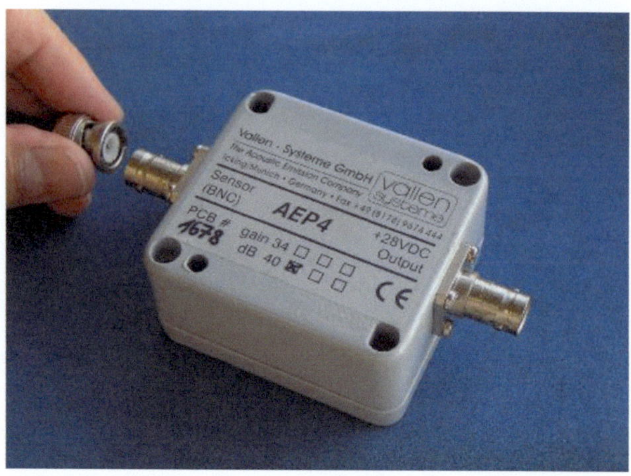

capture electric noise. If any intrusion of noise is recognized, the preamplifier must be isolated from the source of the noise, or measures must be taken to counteract electromagnetic waves.

5.1.3 Sensitivity Check of an AE Sensor

After the complete setup of the devices and installation of the sensor, one must confirm that these tasks have been properly performed. The sensor's sensitivity, including the effects of installation of the sensor and signal cables, must also be checked.

5.1.3.1 Review of the Overall Connection Status

To confirm that the entire signal line is properly connected, a measurement object is lightly hit, as shown in Fig. 5.11, or artificial AE signals are generated with a pulse generator, thereby checking that signals are being input into the sensor for each channel. If a channel gives no response, one should check the set values of the AE devices, the installation condition of the sensor, the connections of the signal cables and preamplifier, and the possible failure of individual devices. Appropriate measures must be taken to resolve any problem that is noted.

5.1.3.2 Sensitivity Check by Pencil Lead Break

The sensitivity of each sensor must be checked after confirming the status of the overall connection.

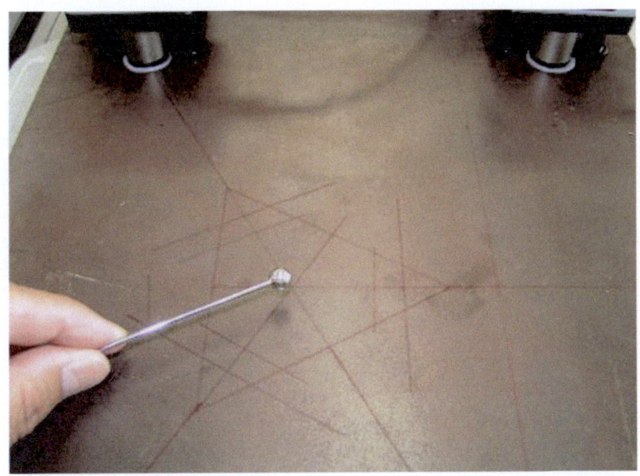

Fig. 5.11 Check of the connection status

A simple method for an on-site sensitivity check is to use the lead of a mechanical pencil that is pressed until broken. In this method, the AE wave that is generated when the lead of the pencil is broken is treated as a AE source to check the amplitude of the signal input into the sensor. In this sensitivity check, it is determined whether each sensor demonstrates a given sensitivity, and whether any difference in sensitivity between sensors falls within a given range. In general, sensitivity is checked in terms of an average amplitude determined for the same sensor, after several repetitions of breaking the pencil lead (Hsu-Nielsen source), in consideration of variations in the intensities of AE waves generated. In this process, because the amplitude obtained varies with the material and thickness of the measurement object, one should conduct a preliminary test to confirm the amplitudes to be obtained. For this sensitivity check, the position and angle at which the pencil lead is broken, as well as the thickness and type of lead, should be kept constant, for the purpose of reproducibility. The adaptor shown in Fig. 5.12 ensures that the press/breakage angle of the pencil is constant, and it thus improves reproducibility.

5.1.3.3 Sensitivity Check by a Pulse Generator

The previous section described a method that uses a breaking pencil lead as a AE source. When higher reproducibility is required, a pulse generator (pulsar) is used as a AE source to check sensitivity. In this method, an artificial AE generated by the pulse generator is used as a AE source to confirm the amplitude of the signal input into the sensor (refer to Fig. 5.13).

In this method, the amplitude of the artificial AE may vary with the position of the pulse generator and the pressure imparted to the pulse generator. For this reason, one should keep the interval between the sensor and pulse generator constant, and keep the pressure imparted to the pulse generator constant by means of a jig. In

Fig. 5.12 A pencil breaking apparatus

Fig. 5.13 Sensitivity check by a pulse generator (*Left*: pulse generator, *Right*: sensor)

addition, the most recent device has a function that can automatically measure the sensitivity and propagation time of an adjacent sensor, and calculate the sensor-to-sensor interval and wave velocity using a measuring sensor as a pulse generator.

5.2   Noise

Yoshihiro Mizutani

In AE testing, a very weak signal with magnitude of several tens of microvolts to several tens of millivolts is amplified by a high-gain amplifier to measure. It is important to create a noiseless environment for the AE measurement, as in the same way that the chattering of birds and the sound of swaying plants can be recognized

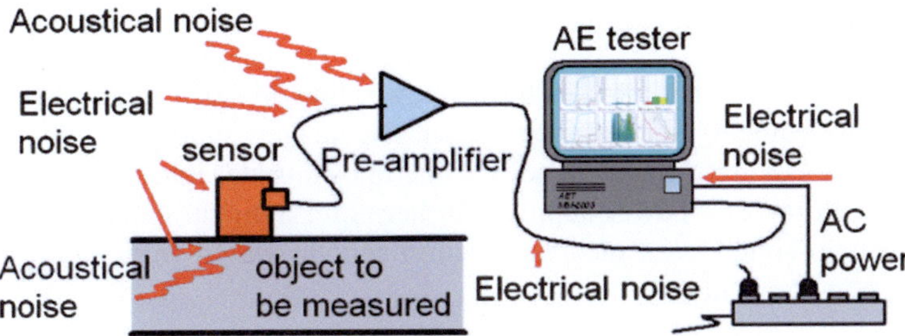

Fig. 5.14 Some noises observed during an AE test

only in a quiet environment. Problematic noise includes not only acoustic noise from mechanical vibrations, but also electric noise. Without sufficient knowledge of noise, the low-noise environment required for AE measurement cannot be realized. The ratio of the signal to noise is referred to as the SN ratio. Realizing a high SN ratio is important when conducting AE testing. The noise level is sometimes expressed using the ASL (average signal level) or RMS (root-mean-square), which were described in Chap. 4. This section presents types of noise that are problematic during AE measurements, and describes preventive measures and countermeasures against noise.

5.2.1   Types of Noise

"Electric noise" and "acoustic noise" as shown schematically in Fig. 5.14 intrude in various forms during AE testing. It is necessary to eliminate noise that interferes with AE measurement.

Two typical types of noise that are problematic during AE measurement are presented below.

5.2.1.1   Acoustic Noise

Acoustic noise is noise that results mainly from mechanical vibrations, and is elastic waves generated in an object to be measured, but not AE waves that should be measured. When the object to be measured is a rotary device, or if the object is connected to a vibration source via piping, a vibration noise may intrude. In a materials test, the frictional noise of a pin or a chuck, used to apply loads to a test specimen, can become problems in some cases.

5.2.1.2   Electric Noise

Electric noise is noise that intrudes from a power line or signal cable. When an AE measuring instrument is connected to an AC power supply to which noise-generating devices are connected, noise sometimes intrudes into the AE measuring instrument via the AC power supply. Furthermore, a signal cable serving as an antenna can in some cases pick up electromagnetic waves from a broadcasting station or power machinery.

5.2.2   Preventive Measures and Countermeasures Against Generated Noise

The previous section presented two general types of noise. This section describes preventive measures against the generation of electric and acoustic noise, as well as countermeasures against noise.

1. Acoustic noise

When the source of acoustic noise can be predicted, a sensor must be placed in a position where the sensor is insusceptible to noise. Furthermore, an AE sensor with a frequency characteristic different from that of noise should be selected, with consideration of the frequency band of the AE signal desired for detection. When acoustic noise is not eliminated even by these measures, the noise is eliminated by frequency filters, such as a high-pass filter or bandpass filter. When an object to be measured vibrates, the signal cable should be fixed such that it does not vibrate with the objects.

When the time range for AE occurrences can be predicted, as in the case that AE signals are generated due to fatigue crack propagation during a fatigue test, sometimes only signals within a specific time range are detected by a strobe function (that inputs external signals and measures only AE signals within a specific time range), as shown in Fig 5.15. In another method, guard sensors, as described in Sect. 6.2, in Chap. 2, are used to eliminate acoustic noise propagating from outside the range of interest.

5.2.2.1   Electric Noise

A primary preventive method to protect against the intrusion of electric noise is to use the manufacturer's genuine devices, sensors and cables as specified. When electric noise coming from sources (a) to (c) described below is detected, the following countermeasures should be considered even after the above method has been used.

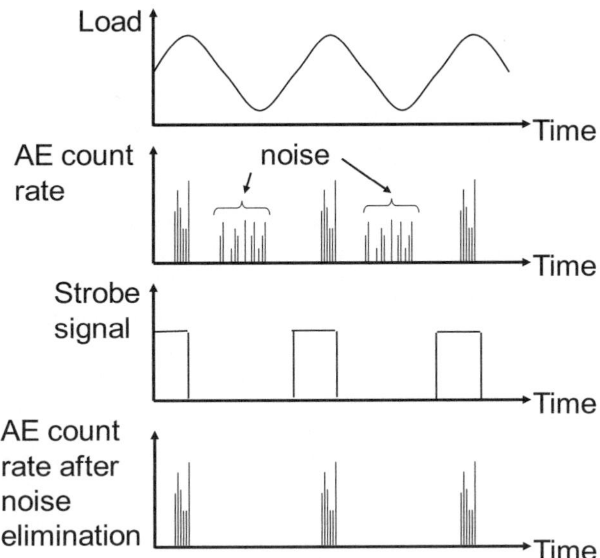

Fig. 5.15 Noise elimination using a strobe function

(a) Noise intrusion via a power cable

AC power is supplied from source that is not used by other devices.

A power line filter or noise filter transformer (noise cut transformer) with noise elimination capability should be used when above countermeasure is not sufficient.

(b) Noise propagation through air, such as electromagnetic waves from a broadcasting station or power equipment

When a device near the sensor generates electromagnetic waves, the device must be turned off or moved away from AE systems. Since a signal cable may become an antenna for electromagnetic waves, noise can intrude from the cable. For this reason, a short signal cable should be used and the looping of the signal cable should be minimized.

A sensor with a built-in preamplifier, as described in Sect. 3.5, Chap. 3, can almost completely cut-out noise intruding between the AE sensor and the preamplifier.

(c) Noise from ground currents

To keep a ground current from passing between the AE sensor ground and the instrument's ground, the use of only one grounding point is recommended. In general, because of the difficulty of insulating the instrument from the ground, the sensor is insulated from the ground. In particular, when a longer cable is used in an on-site test, the countermeasures described above may need to be taken.

5.2.3 *Identification of the Noise Source*

If noise is measured in an actual test, the source of the noise must be identified and be eliminated employing the abovementioned methods. Noise monitoring is basically conducted under the AE testing conditions that are actually used.

Since it is difficult to identify a noise source only from AE parameters, AE waveforms are used when an AE instrument has such a function. When an AE instrument does not have this function, a digital oscilloscope or other instruments are used to check waveforms. If a user becomes familiar with AE testing, he or she can sometimes evaluate types of noise by connecting an audio monitor, as described in Sect. 4.2 and 4.4, Chap. 4. In any case, much experience is required to determine whether noise is acoustic or electric, and to diagnose the cause of the noise. Detailed steps for identifying noise sources are given below.

1. Display AE waveforms or use an audio monitor to examine the frequencies and duration of noise signals to predict types of noise.
2. Supply AC power from another source. Use a power line filter or noise filter transformer.
3. Ground the device differently.
4. Replace a long signal cable with a short one. Check whether the cable forms a loop or not.
5. Cut-off suspected propagation routes of acoustic noise.
6. Prevent vibration from sources such as motors and other power machinery. Check whether the cable vibrates or not.
7. Turn off power sources that could generate electromagnetic waves.
8. Remove the sensor from the test specimen. (Electrically insulate the sensor and the object to be measured.)
9. Replace an existing signal cable with a new one. (Assume that there is a defective signal cable.)

5.3 Test Methods

Shigenori Yuyama

5.3.1 *Loading Method*

5.3.1.1 Material Evaluation Test

Material tests, such as tensile, bending, fracture toughness, and fatigue tests, are normally conducted to evaluate the strength and characteristics of materials. AE

Fig. 5.16 Tensile test

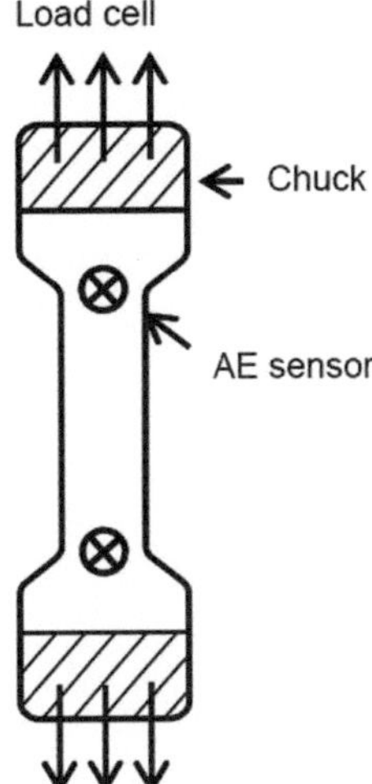

Fig. 5.17 Bending test

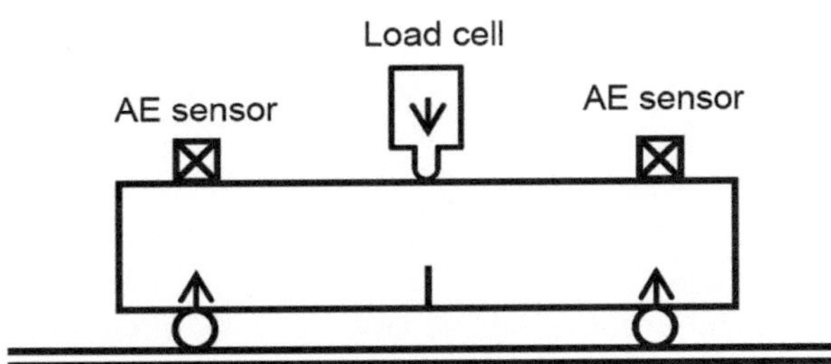

measurement is employed to evaluate deformation characteristics, micro-crack generation, or crack growth history. Loading methods and AE sensor locations on test specimens are schematically shown in Figs. 5.16, 5.17, and 5.18.

Figure 5.16 shows a loading method employed during a tensile test. For the tensile test, a rectangular or round-bar specimen is generally used. Both ends of the specimen are arranged to be well fixed by a chuck, to avoid slippage between the chuck and specimen. A load is then applied to the specimen at a proper rate. The AE measurement is conducted to detect generation of AE associated with plastic or twinning deformation or martensitic transformation during the test. In most cases, it is sufficient to analyze a signal detected by one sensor. When noise resulting from slippage between the specimen and chuck is significant, a spatial filter based on the linear source location using two AE sensors placed at the ends of the specimen is applied to eliminate extraneous noise.

Figure 5.17 shows a loading method used during a three-point bending test. This bending test is often used to evaluate the strength and toughness of concrete specimens. For this test, two AE sensors are usually placed near the two ends of the specimen, and a spatial filter is applied to eliminate noise. After the noise has been eliminated, only effective signals are used for analysis.

Fig. 5.18 Fracture
toughness test

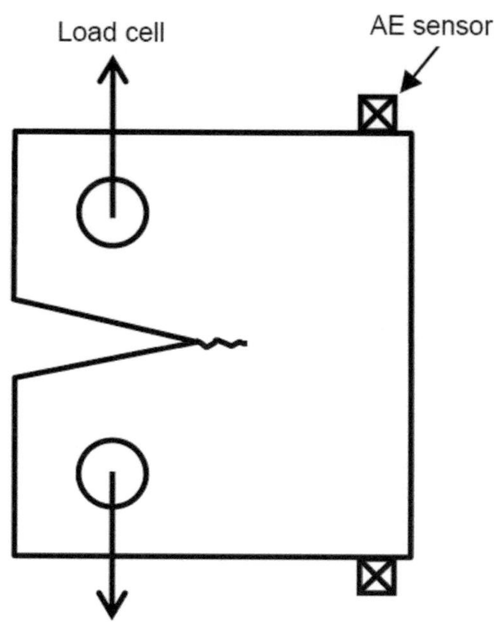

Figure 5.18 illustrates a fracture toughness test using a compact tension specimen. AE sensor positions are schematically shown in the figure.

Normally, two AE sensors are placed at the two ends of the test specimen to
detect and evaluate fatigue crack growth taking place at the tip of the fatigued
pre-crack.

Although fatigue tests are conducted under repeated loading, basic methods for
the loading and installation of AE sensors are the same as those used in tensile,
bending and fracture toughness tests. Because the load varies continuously under
repeated loading, mechanical noise is often generated continuously at the chuck and
pin supporting the load. Therefore, a spatial filter is applied to eliminate noise.

5.3.1.2   AE Testing of Structure

To evaluate structural integrity, AE testing is employed for various types of metal,
composite, or concrete structures such as refineries, chemical plants, power stations, offshore rigs, bridges, tunnels, buildings, aircraft, and rock structures.

AE testing is carried out in two ways in the field, namely pre-service and
in-service inspections. Structures are loaded pneumatically or hydrostatically in
accordance with a predetermined load schedule in either pre-service or in-service
inspection during periodic shut-down. Testing is schematically illustrated in
Fig. 5.19. When significant signals that continuously increase under stimulation
are produced or intense AE sources are detected during pneumatic or hydrostatic

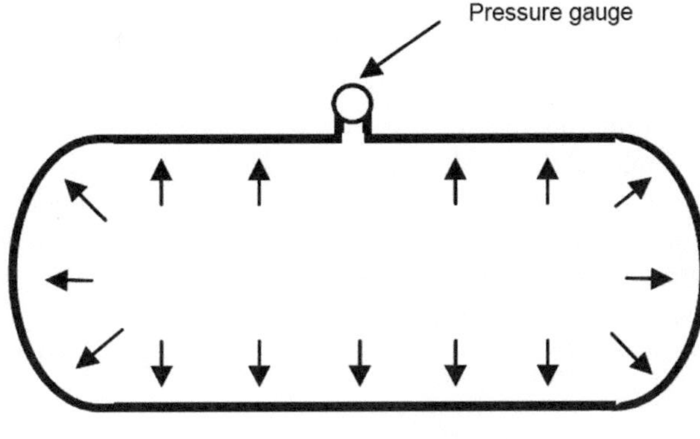

Fig. 5.19 Hydrostatic test of a pressure vessel

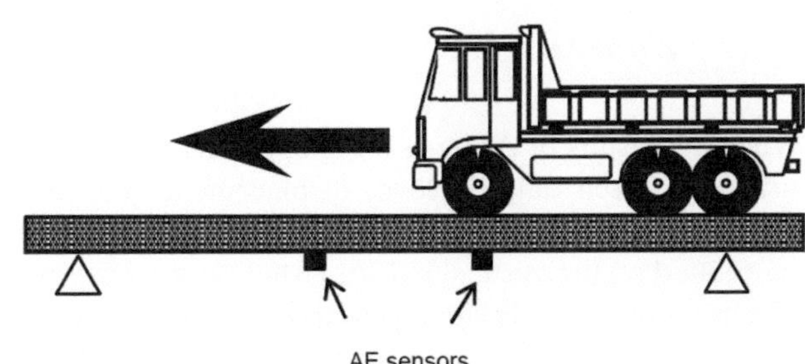

Fig. 5.20 AE test of a reinforced concrete beam by mobile loads

tests, the AE examiner reports the finding to the test manager for proper action to be taken.

To evaluate the structural integrity of bridges, heavy vehicles with different loads are often driven repeatedly across the structure. This situation is illustrated in Fig. 5.20. If a structure is in operation, AE data are analyzed according to the load variation resulting from the operation. For example, AE analysis is often carried out during the start-up or shut-down in chemical plants or refineries. In the case of bridges, AE activity is observed under live loads due to traffic. In the case of railway bridges, AE activity in response to load changes due to passing trains is commonly analyzed.

5.3.2   Data Sampling of External Analog Parameters

In materials tests or structural tests, external analog data such as load, strain, and pressure data are input into the AE instrument in conjunction with the AE data recorded during the tests. Such analog data are input into the instrument as analog signals of ± 1 to ± 10 V, digitalized, and recorded as analog parameter data along with the AE data. In material tests, the sampling rate is usually set to about 1 sample/s. In an impact fracture test, where a fracture is rapidly induced, or a fatigue test, where the relationship between the loading phase and AE activity under cyclic loading is analyzed precisely, the rate is set to 10 to 100 samples/s. When long-term measurement such as stress corrosion cracking or creep tests is conducted over several weeks to several months, the rate is sometimes set to 1 sample per 10 s to 1 min. In structural tests, data sampling is also conducted in the same manner as in material tests. During AE tests of pressure components, analog data from a pressure transducer is acquired by an AE instrument together with AE data. In bridge monitoring, displacement data indicating the deflection of the bridge and the strain of reinforced bars are sampled as external parameters. When analog signals cannot be input directly into the AE instrument during testing, analog signals may be manually input into the instrument by a potentiometer.

5.3.3 Data Display

When the results of AE tests are graphically presented for laboratory specimens or structures, such data as load, displacement and strain data are plotted on the horizontal axis, and the AE data (e.g., the AE signal amplitude, count, or energy) are plotted on the vertical axis to visualize and analyze crack initiation and growth that has produced AE signals. In tensile, bending, and fracture toughness tests, the histories of AE activity are often plotted with load and displacement on the same graph. In long-term tests such as stress corrosion cracking and creep tests, the elapsed time is again used on the horizontal axis in comparison to the AE activity. Examples of basic data displays are given in Fig. 5.21. Analysis software normally has graphical functions such as the bar graph, correlation plot, line graph and stair graph as standard selections. Bar graphs are often used to compare AE data with other parameters. Correlation plots are employed to investigate correlation among AE parameters. Line and stair graphs are used to observe the historical behavior of AE activity over other parameters. Examples of a correlation plot and stair graph

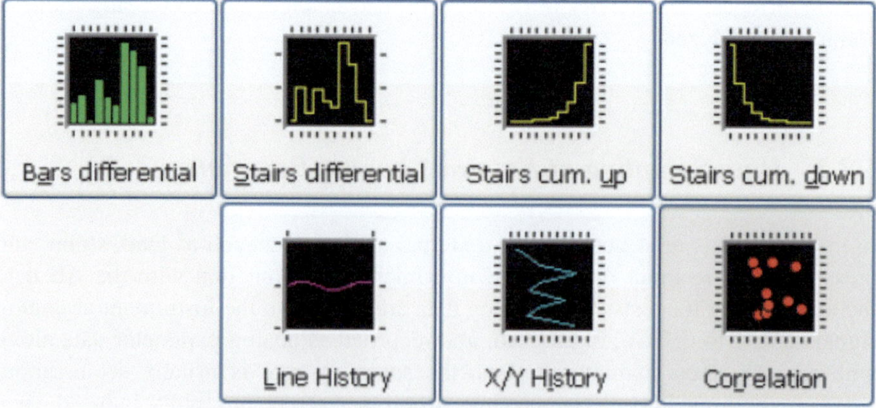

Fig. 5.21 Sub-menu for graph setting

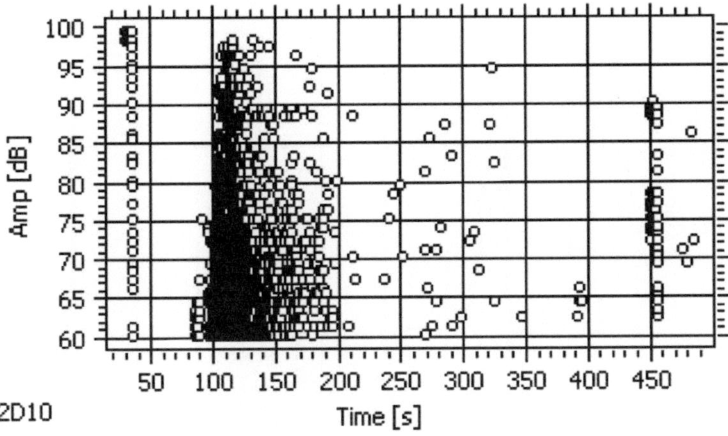

Fig. 5.22 Correlation plot

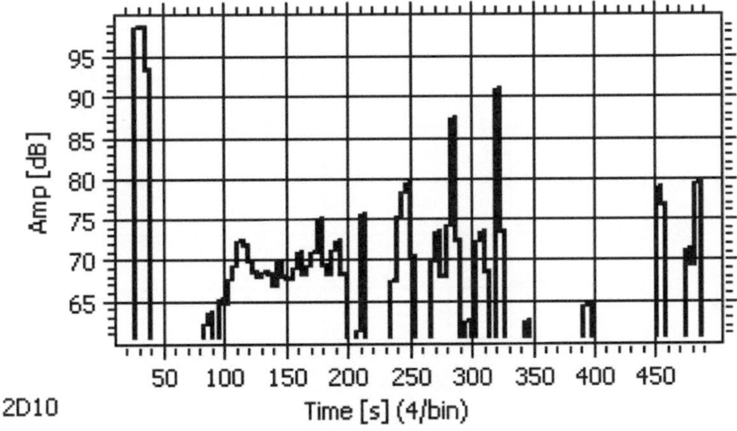

Fig. 5.23 Stair differential chart

exhibiting the relationship between amplitude (in decibel scale) and elapsed time
are shown in Figs. 5.22 and 5.23, respectively. Average values of the data in bins of
4 s are plotted as a function of the elapsed time. Any bin width can be set by the
operator. A stair graph can also present the maximum and minimum values
(Max-Min), the total (Sum), and the maximum of the data within the bin (Maximum) instead of the average. Figure 5.24 shows an example of a stair graph for the
total (Sum) of data in each bin of 4 s as a function of the elapsed time.

Figures 5.25 and 5.26 show the history of cumulative hits, selecting elapsed time
for the horizontal axis and the number of hits for the vertical axis. Figure 5.25 is a
stair graph for the cumulative number of hits in increasing mode, while Fig. 5.26 is
a stair graph for the cumulative number of hits in decreasing mode. Note that the

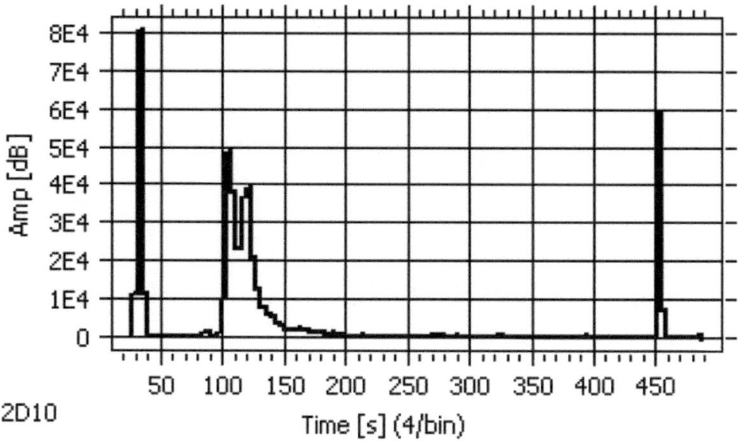

Fig. 5.24 Stair differential chart by bin summation

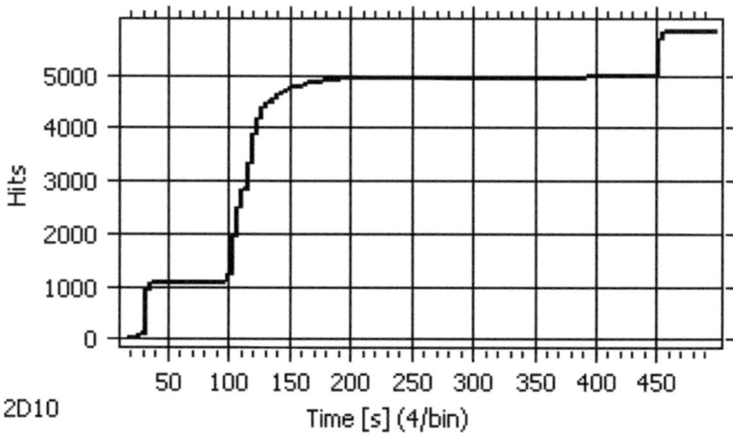

Fig. 5.25 Cumulative AE hits (minimum to maximum)

former version of the plot (Fig. 5.25) is usually used. Three parameters can be displayed together in a three-dimensional display. Figure 5.27 shows the elapsed time (X axis), number of hits (Y axis), and AE signal amplitude (Z axis) as a three-dimensional display. In actual data analysis, AE data and external parameters (load, displacement, and strain) acquired simultaneously are often displayed on the same graph. For instance, Fig. 5.28 shows AE activity during a fracture toughness test of a high-strength aluminum alloy, and Fig. 5.29 shows AE activity observed during a three-point bending test of a concrete beam reinforced with a carbon-fiber-reinforced plastic panel. These figures clearly indicate that a crack initiates or an

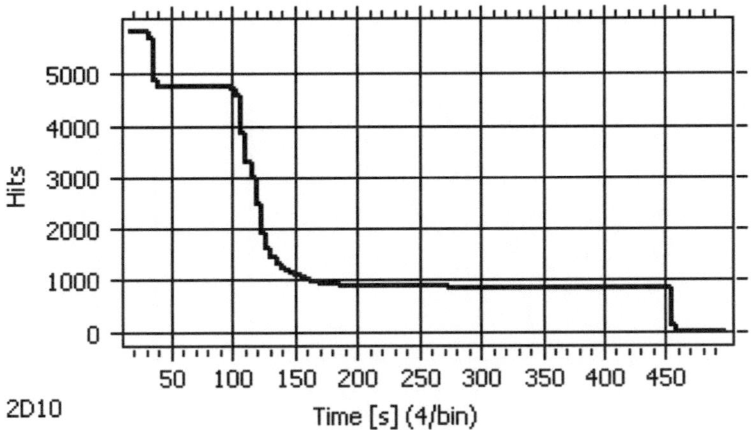

Fig. 5.26 Cumulative AE hits (maximum to minimum)

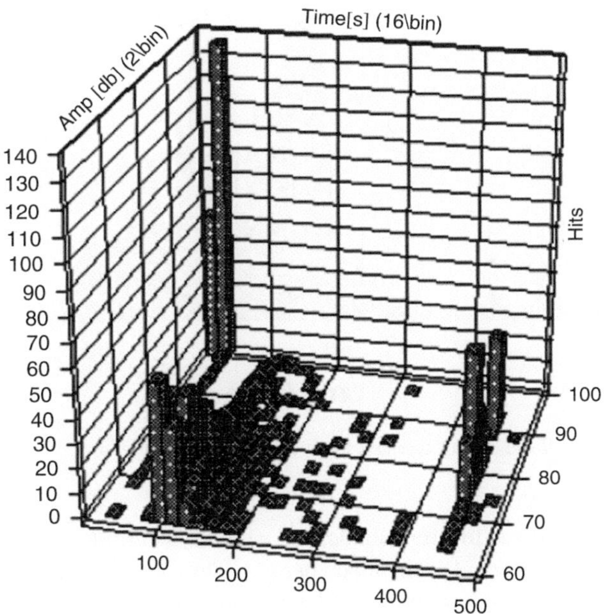

Fig. 5.27 3D display of time (X axis), number of hits (Y axis), and AE signal amplitude (Z axis)

existing crack has begun to grow at the point where AE activity increases significantly in response to a change in load and other parameters.

Figure 5.30 presents the historical AE activity and strain change observed in a structural test of a concrete bridge under repeated loading due to the passing of a

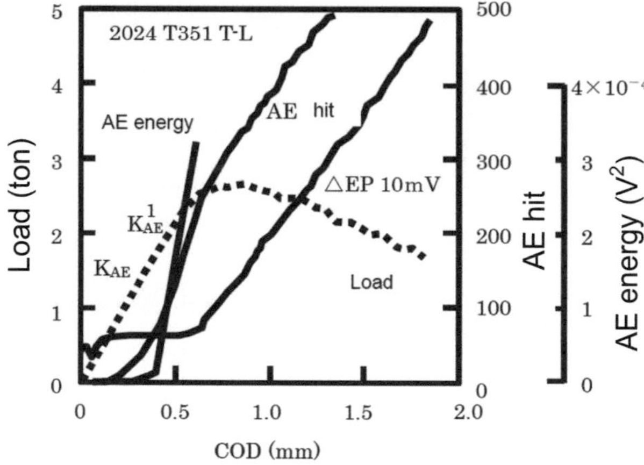

Fig. 5.28 AE characteristics in a fracture toughness test of aluminum alloy 2024

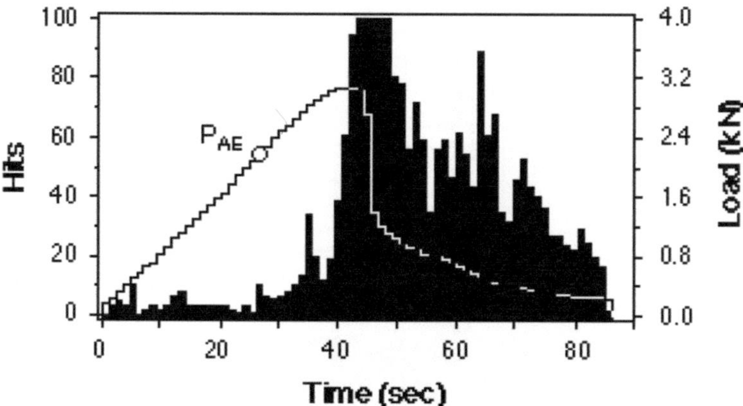

Fig. 5.29 Histories of the AE hit rate and load; P_{AE} is the load at which AE activity remarkably increases

dump truck with three different loads. AE signals resulting from friction within existing cracks are detected as the load changes. The number of detected AE signals depends on the load of the truck. It has been demonstrated that the level of damage to the concrete beam can be evaluated by observing the AE activity associated with the load change. In AE data analysis, amplitude distributions, as illustrated in Figs. 5.37 and 5.38, are often used to eliminate noise and identify AE sources. These figures show the relationship between the detected signal amplitudes (Am) and the number (n) of signals (hits). Details of the amplitude distribution are described in the next section.

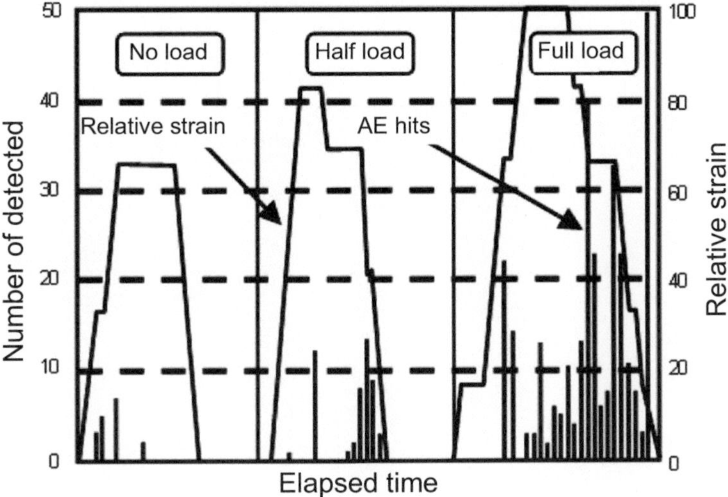

Fig. 5.30 Histories of the AE hit rate and strain on the main reinforcement of a concrete bridge under repeated loading due to the passing of a dump truck

5.4 Evaluation of Measurement Data and Test Recordings

Tomoki Shiotani

5.4.1 Interpretation and Evaluation of Measurement Data

It is preferred to record all AE signals exceeding a threshold as AE waveforms. However, it is not easy to record all burst (transient) AE signals with current sampling/recording speeds and capacities of recording media. For this reason, instead of recording AE waveforms, AE parametric features characterizing AE signal waveforms, as explained in Sect. 2.5 of Chap. 2, are generally recorded to evaluate materials. In this section, taking the example of a uniaxial compression test of a cylindrical concrete specimen, as illustrated in Fig. 5.31, material evaluation based on AE parameters as well as AE sources determined in the test is demonstrated.

5.4.1.1 Test Conditions

Figure 5.31 outlines the test. The cylindrical specimen has a diameter of 10 cm and height of 20 cm, and a total of six AE sensors are installed on the upper and lower surfaces at intervals of 120° for the measurement.

The data recorded in this test are the histories of load/stress data and AE parameters. Hard sponge is used in the compression test to prevent noise due to contact between the

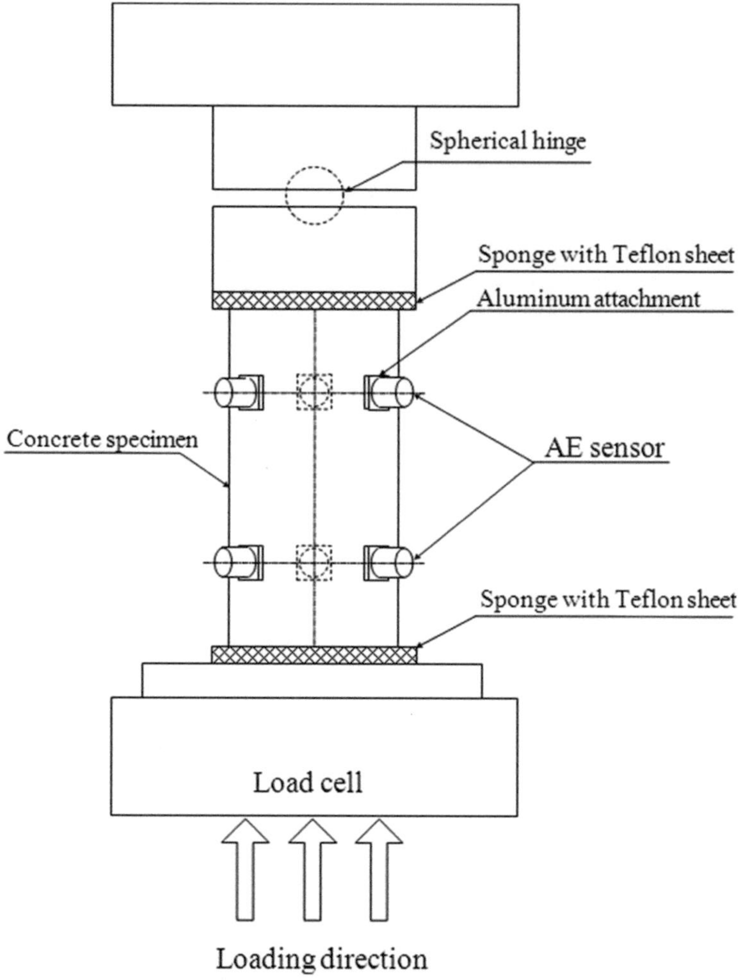

Fig. 5.31 Uniaxial compression test of concrete

end face of the specimen and the loading plate during the test. The sponge has the additional role to fill any space at the interface. Hard sponge or rubber is also used to place at the loading points or supporting points in three-point/four-point bending tests of prism-shaped specimens to eliminate mechanical noise due to contact.

5.4.1.2   Evaluation of Data Series

AE data are generally evaluated over time or for a varying applied load. As an example of AE data evaluation, Fig. 5.32 shows the time history of AE hits obtained from all six sensors, while Fig. 5.33 shows the time history of AE energy obtained from all six sensors.

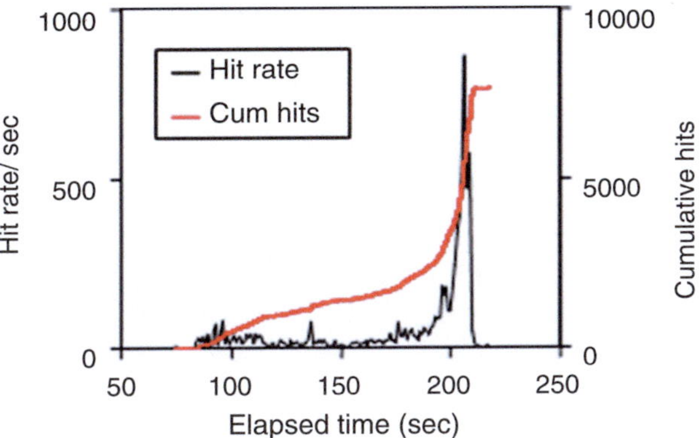

Fig. 5.32 AE hits versus time

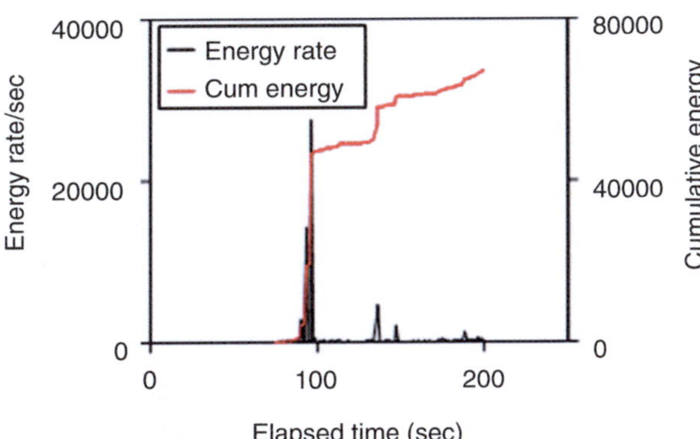

Fig. 5.33 AE energy versus time

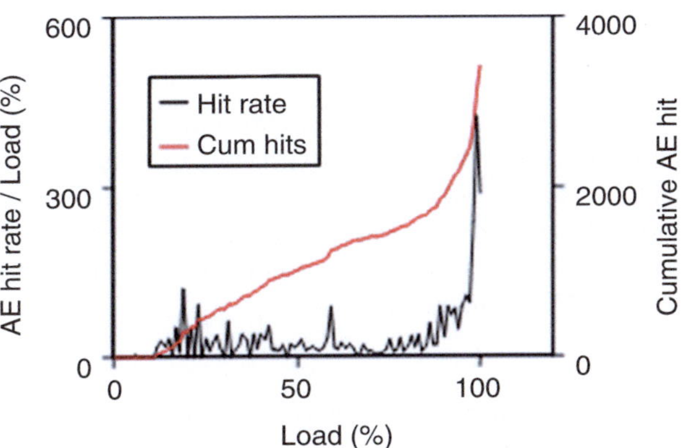

Fig. 5.34 AE hits versus load

It is seen that AE activity increases as approaching to the ultimate stage of loading in the AE hit-based evaluation, while there is a high level of AE energy in the initial stage of loading in the energy-based evaluation.

Figures 5.34 and 5.35 show the load histories for AE hits and AE energy. The load on the horizontal axis is normalized by the maximum value obtained in the test i.e., it is a percentage of the maximum load. The use of the normalized load history

Fig. 5.35 AE energy versus load

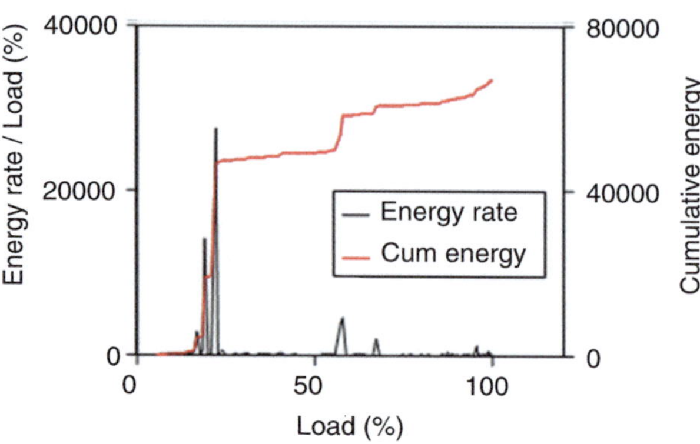

Fig. 5.36 Correlation between energy and duration

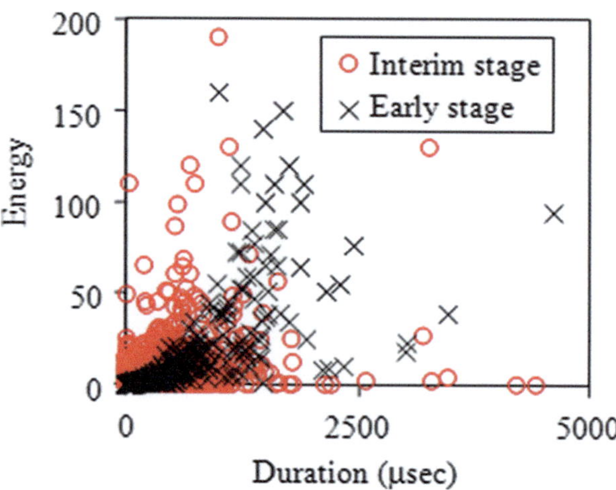

makes it possible to evaluate parameters such as the intensity of AE activity, which may increase as the number of AE hits increases from 90 % of the maximum load (see Fig. 5.34), and the fracture scale, which appears to be more significant at 20 % of the maximum load (see Fig. 5.35) according to energy. In this way, the fracture process of materials can be examined and evaluated by studying the histories of various AE parameters.

5.4.1.3 Evaluation of Correlations Among AE Parameters

To examine the factors contributing to AE sources generated by crack nucleation or growth due to tensile or shear deformation or friction within defects, the evaluation of correlations among AE parameters is effective in some cases in addition to the quantitative analysis of waveforms. As an example, a chart showing the correlation between AE energy and its duration is plotted in Fig. 5.36. The figure shows that the AE signals generated in the initial stage of loading and after the intermediate stage of loading represent respective characteristics between duration and energy. For instance, AE signals obtained after the intermediate stage are found to have high energy accompanied with short duration. In this way, the consideration of correlations among AE parameters allows examination of the scale and mechanism of the fracture resulting in AE sources.

5.4.1.4   Evaluation of Frequency in AE Parameters

The proper evaluation of the frequency change in AE parameters may allow assessment of the process of fracture. A representative evaluation of the frequnecy in AE parameters is based on the AE peak amplitude, which is referred to as the amplitude distribution. Figure 5.37 shows an example of an amplitude distribution. The bars in the figure (left vertical axis) represent the occurrence frequencies of amplitudes (in 1 dB bins) obtained from AE measurement. This distribution is referred to as the differential amplitude distribution. The figure shows that there are few AE signals with large AE amplitude, while there are many AE signals with small amplitude. Focusing on the gradient and variations of this amplitude, by characterizing the amplitude distribution, several studies have been conducted, for example, to distinguish AE activity from noise or to identify factors contributing to the generation of AE signals.

In practice, the distribution of the cumulative amplitude (cumulative amplitude distribution, see the solid line in Fig. 5.37 and the right vertical axis), which is obtained from the cumulative number of AE frequencies exceeding each amplitude, is logarithmically expressed and used for actual analysis, instead of the bar charts also shown in Fig. 5.37. The negative gradient of the curves is referred to as the b-value[1]. In Fig. 5.38, AE data from Fig. 5.37 are divided into data for the initial load and for the intermediate load to draw individual amplitude distributions. In Fig. 5.38, the bar charts and solid lines represent the differential amplitude distribution and distribution of cumulative amplitudes, respectively, in the same way as in Fig. 5.37. In the figure, two distributions—one obtained for the initial stage of loading and other after the intermediate stage of loading—are drawn. From the figure, it is estimated that the gradient of the cumulative amplitude distribution for the initial stage of loading is less than that after the intermediate stage of loading,

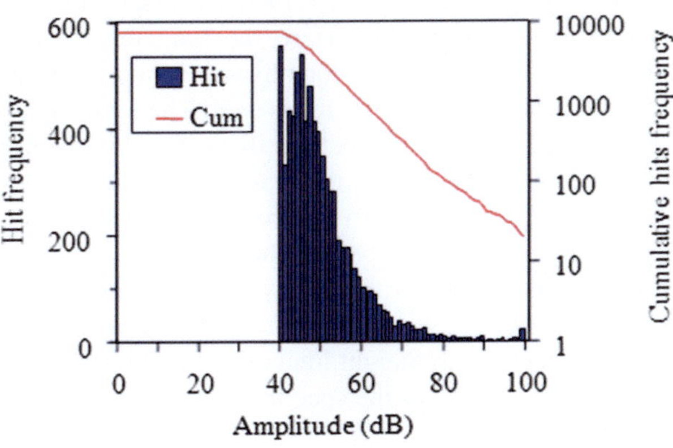

Fig. 5.37 Amplitude distribution in a concrete uniaxial compression test

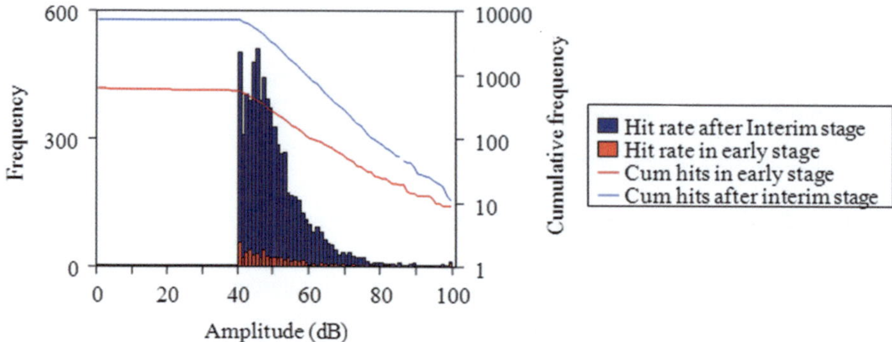

Fig. 5.38 Amplitude distributions with progression of damage

suggesting that AE activity with large amplitude occurs more frequently relative to AE activity with small amplitude during initial loading than that in the case after the intermediate stage.

5.4.1.5 Evaluation Based on AE Source Locations

As described in Sect. 2.6 of Chap. 2, the identification of an AE source is referred to as AE source location. There are three ways to implement source location based on the number of AE sensors, arrays and calculation algorithms, namely: one-dimensional (1D), two-dimensional (2D), and three-dimensional (3D) source locations. The combination of source locations and AE parameters allows us to determine when and where the AE signal was generated, and what type of AE occurred. This section presents evaluation methods based on 1D, 2D and 3D source locations obtained during the uniaxial compression test of concrete specimen described earlier.

(a) One-dimensional (Linear) source location

By considering AE source locations with load histories, one can evaluate when an AE occurred and which type of AE it was. Figure 5.39 plots the heights of 1D AE sources with elapsing time. Here, the diameter of each circle is drawn to reflect the number of AE counts. The load history is superimposed on the AE events (see the right vertical axis in the figure) as well. The figure shows that an AE event occurred near a height of 100 mm before 100 s, the AE event gradually moved toward the bottom of the specimen in the latter half of the fracture process, and the AE event with a large parameter (here AE count) was observed mainly in the lower area of the specimen immediately before the fracture. The intensity of spatial AE source distributions can be evaluated by accumulating the frequency of 1D AE sources for each height of the event. Figure 5.40 is a bar chart in which the accumulated AE events in each 5-mm height interval are exhibited. The vertical and horizontal axes represent the height of the specimen and the accumulated frequency of AE events,

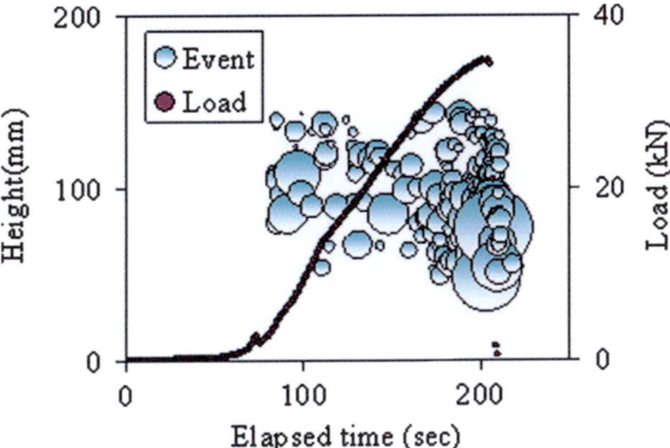

Fig. 5.39 Time history of AE events (with the diameter of a circle reflecting the AE count)

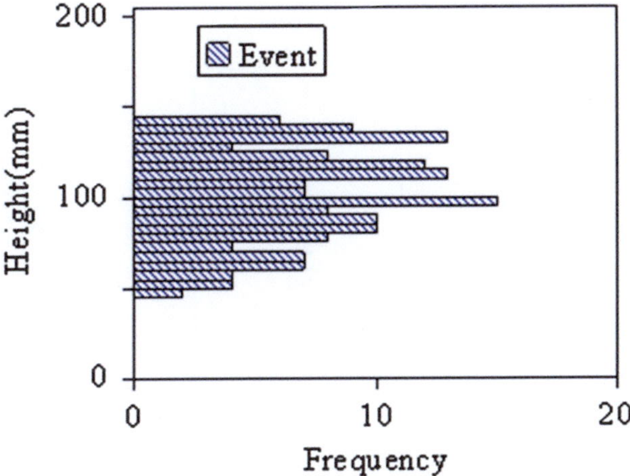

Fig. 5.40 AE events by unit height

respectively. From the figure, one can estimate that the AE activity in the upper area of the specimen (100–150 mm) was more intense than that in the lower area (50–100 mm). Here, it is also possible to evaluate the AE event frequencies on the horizontal axes using the various AE parameters presented instead of AE events. Furthermore, drawing the same distribution of frequencies for each fracture step as specified by an engineer is also helpful in associating AE data with fracture phenomena.

(b) **Two dimensional (Planar) source location**

2D source location is effective in evaluating the characteristics of planar AE events. Figure 5.41 shows 2D AE sources. In the figure, the horizontal and vertical axes represent width and height, respectively. The positions of AE sensors are denoted by "+". A concentration of AE events from the upper left to the lower right of the specimen can be observed. Figure 5.42 shows AE events weighted by an AE

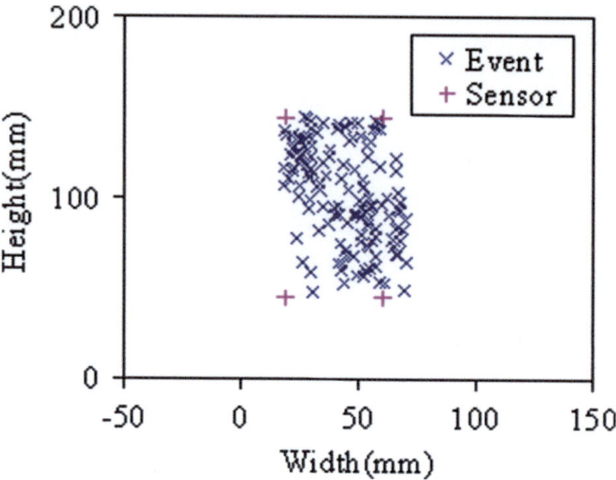

Fig. 5.41 3D AE events

Fig. 5.42 2D AE events weighted by an AE parameter (with the diameter of a circle reflecting the AE count)

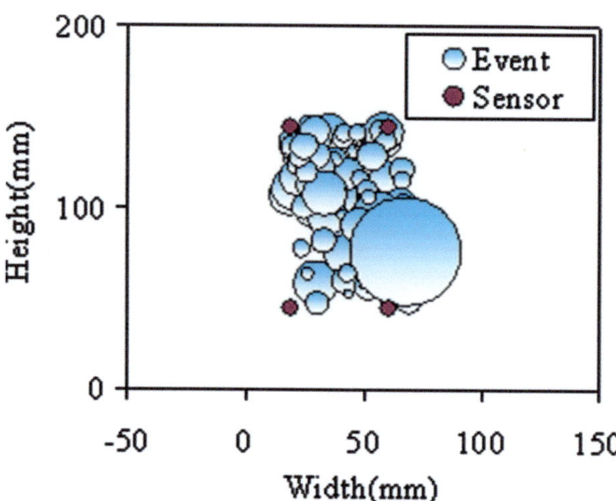

parameter, namely AE count. This chart allows to determine where an AE event occurred and how large it was. In addition, by drawing the same type of figure for unit time or unit fracture phase, the spatial progress of AE events in each fracture phase can be evaluated.

(c) Three dimensional source location

3D source location determines the spatial coordinates of AE events, which allows engineers to draw unique AE charts or interpret 3D AE data according to their experience and insights. Figure 5.43 shows 3D plots projected on the plane, front and side of the specimen. In all plots, the position of an AE sensor is denoted by "+". AE events are observed at the center of the plane and on the front of the specimen in Fig. 5.43c, a, and near the front of the specimen in Fig. 5.43b.

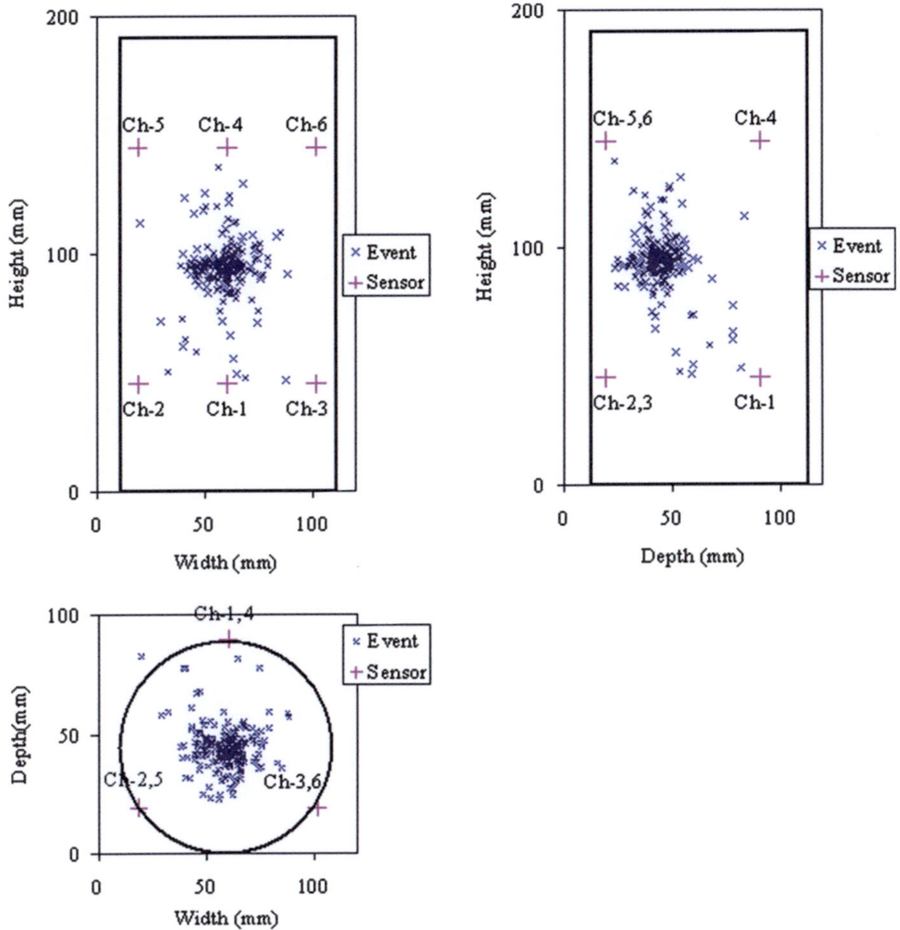

Fig. 5.43   Visualization of 3D AE events

Figure 5.44 shows 3D plots of source location in which the AE counts of an AE event are represented by the diameter of a circular. From Fig. 5.44b, it is found that AE events observed at the intermediate height of the specimen, particularly those near the front, have a large AE count. For reference, a case in which the source locations are plotted in 3D coordinates is presented in Fig. 5.45. Recently, analysis based on this 3D display has been well conducted.

5.4.1.6   Criteria for Structural Integrity in AE Testing

As an example of the evaluation criteria necessary for structural maintenance, Fig. 5.46 depicts a criterion for classifying damage in concrete (see NDIS2421 or

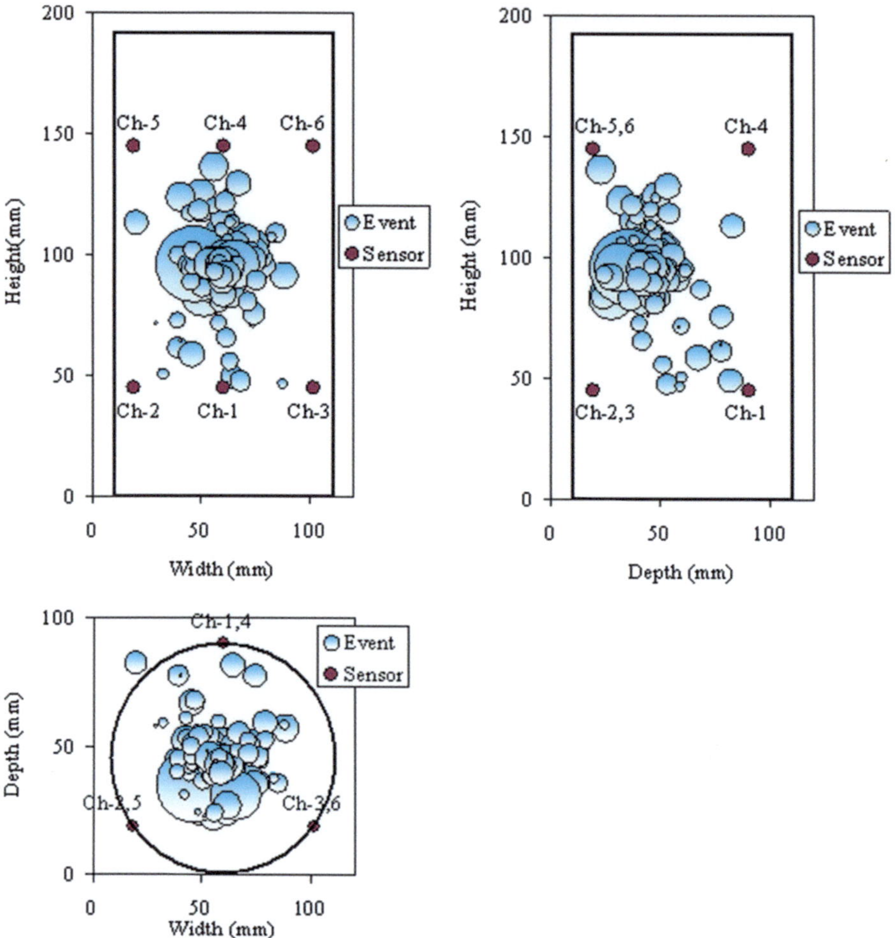

Fig. 5.44 Visualization of 3D AE events (with the diameter of a circle reflecting the AE count)

Recommendation of RILEM TC-212-ACD). The figure is composed of two promising parameters: Calm and Load, and these indices are obtained during the repeated load applications of concrete. The Calm is a ratio of cumulative AE hits during unloading to that during both loading and unloading, and the Load is a ratio of the value of such reference parameters as deformation, strain, and load showing the onset of AE activity to the past maximum value of the parameter. As combining these two parameters as shown in this chart, one can determine the damage of concrete as serious, intermediate and moderate/ intact condition.

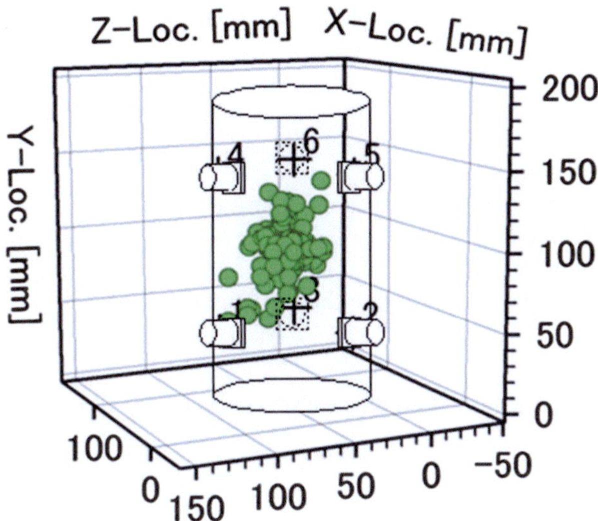

Fig. 5.45 3D image of AE events

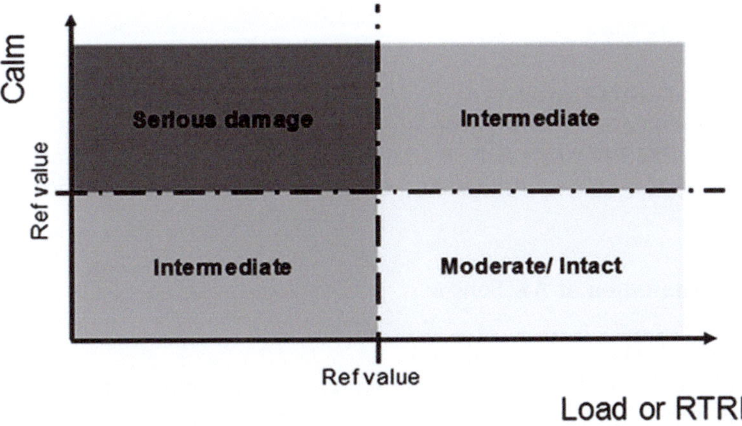

Fig. 5.46 Criterion for classifying damage in concrete

5.4.2   Records of AE Testing

Examples of items recorded in AE testing are given below. It is recommended to record the following issues, depending on the field and type of test.

5.4.2.1 Items Relating to the Test Environment

- Test date
- Test location
- Name of tester
- Test devices and jigs
- Test environment (e.g., noise around the test devices)
- Test (loading) method
- Test time/duration
- Other necessary data concerning other test environments

5.4.2.2 Items Relating to Test Pieces/Specimens

- Materials
- Size and shape
- Nominal number of test pieces or name/type of test specimens
- Other necessary data concerning other test specimens or pieces

5.4.2.3 AE Sensors

- Type of AE sensor
- Manufacturer and serial number
- Resonant frequency or sensitive frequency band assured by manufacturer
- Sensitivity test before installation (pencil lead break/contact method)

5.4.2.4 Installation of AE Sensor

- Positions of AE sensor installation
- Installation methods (e.g., crimping/fixing, vinyl tape/magnet holder)
- Couplant (e.g., high-vacuum grease, thermoplastic resin/machine oil/water)
- Sensitivity test after installation (dB/V, pencil lead break/ pulsar method)

5.4.2.5 AE Measurement

- Measurement block diagram (i.e., diagram of equipment connection)
- Type of AE instrumentation
- Manufacturer and serial number
- Frequency filter (HPF/LPF/BPF)
- Gain during measurement (dB)
- AE threshold during measurement (dB or V)

5.4.2.6   Setting of peripheral devices

- Setting conditions for the data logging device
- Setting conditions for other peripheral devices

5.4.2.7   AE Measurement Environment

- Background noise level (dB or V)
- Type of noise and characteristics of the noise signal (e.g., jig friction/electric pulse noise/mechanical vibration/oil pressure source; periodic/discontinuous)
- Other required data regarding noise

5.4.2.8   Recording and Reporting of Test Results

- Recording of data acquired in AE testing and analysis results
- Classification and reporting of AE test results according to documentation on acceptance criteria

Chapter 6
Field Application Examples of AE Testing

Shigenori Yuyama, Masaaki Nakano, Tomoki Shiotani,
and Sunao Sugimoto

Abstract Field applications of AE testing are introduced, showing such integrity evaluations as vessels, pipelines, transformers, bridges, rock slopes and aircrafts. Specifically evaluation of initial ground/rock stress and identification of leakage in pipes with AE testing are demonstrated.

Keywords Field applications of AT Vessels • Pipeline • Gas storage bottle • Above ground storage tank • Leak • Transformer • Railway bridge • Rock slope • Ground stress • Aircraft

6.1 Large Pressure Vessel

Masaaki Nakano

In this section, a case is presented involving the AE testing of reactors made of steel in a desulfurization unit of a petroleum refinery at restart of the plant after periodic inspection. Table 6.1 lists the main specifications of the tested equipment, or reactor B. At restart, the plant enters a transient state in which both the temperature and pressure rise, and this is a good example for AE testing.

S. Yuyama (✉)
Nippon Physical Acoustics, Ltd., Tokyo, Japan
e-mail: yuyama@pacjapan.com

M. Nakano
Chiyoda Corporation, Yokohama, Japan

T. Shiotani
Kyoto University, Kyoto, Japan
e-mail: shiotani.tomoki.2v@kyoto-u.ac.jp

S. Sugimoto
Japan Aerospace Exploration Agency, Tokyo, Japan
e-mail: sugimoto.sunao@jaxa.jp

© Springer Japan 2016
The Japanese Society for Non-Destructive Inspection, *Practical Acoustic Emission Testing*, DOI 10.1007/978-4-431-55072-3_6

Table 6.1 Specifications of the equipment

Item		Reactor B
Dimensions	O.D. (mm)	3302
	Length (mm)	10,000
	Thickness (mm)	151
Material		1.25 Cr–0.5 Mo
Design Press. (MPa)		9.4
Design Temp. (deg. C)		420

6.1.1 AE Testing Method

This application required simultaneous AE measurements of two reactors (A and B) connected in series, and AE sensors with a total of 32 channels were used, with 16 channels for each reactor. The measurement results for Reactor B are presented here. Since the temperature of the tested equipment reaches several hundred degrees Celsius at a steady state, normal AE sensors could not be used. Consequently, high-temperature waveguide sensor/preamplifier assemblies were installed on the tested equipment, using a special attachment, as shown in Fig. 6.1. Figure 6.2 shows the configuration of the AE sensors for reactor B.

Before the AE testing began, the propagation characteristics of the reactors were measured using an artificial AE wave source. The wave velocity was found to be about 3000 m/s, while the attenuation rate was 2–3 dB/m, which corresponded to attenuation of less than 10 dB for the maximum distance between sensors.

Analog signals for pressure at one point and the temperature at three points were input into the AE measurement system as external parameters, and recorded simultaneously with the AE data. After the restart, an AE measurement was conducted continuously for about 60 h until the process reached a steady state.

6.1.2 Results of AE Testing

Figure 6.3 shows the trends of the number of cumulative AE hits in a representative channel and the pressure and temperature. Figure 6.4 shows the trends of the noise levels for 100 min, mainly in the range of 54–55 h after the start of the measurement, for eight of the 16 channels used for Reactor B.

As seen in Fig. 6.4, a high level of noise was often observed, principally as the result of internal fluid flow and temperature variation in the startup period of the plant. In Fig. 6.3, the number of cumulative AE hits reaches tens of thousands. When the AE measurement is conducted under such conditions, software filtering based on the correlation between the AE parameters is effective for discriminating noise.

Figure 6.5 shows a two-dimensional AE source location map for Reactor B. The heads were omitted because there were few located sources in these areas. The

Fig. 6.1 Installation of a high-temperature AE sensor

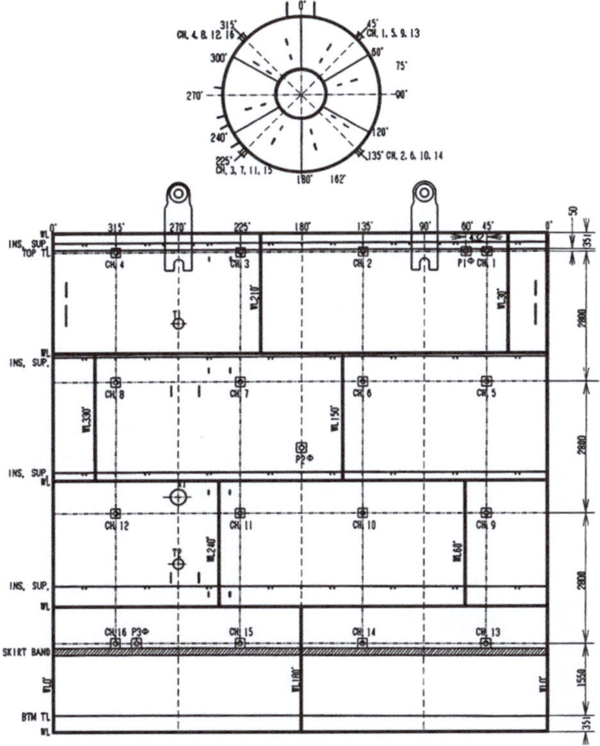

Fig. 6.2 AE sensor configuration (*top head* and shell, Reactor B)

located sources were slightly concentrated in three regions near weld lines, showing clusters; however, these clusters had a small number of AE events and low AE energy. Consequently, they were determined to be Grade C in accordance with the criteria given in Table 6.2. In this test, no AE indicating significant defects was detected in either of the two reactors. The reactors were therefore judged to have no structural integrity problems.

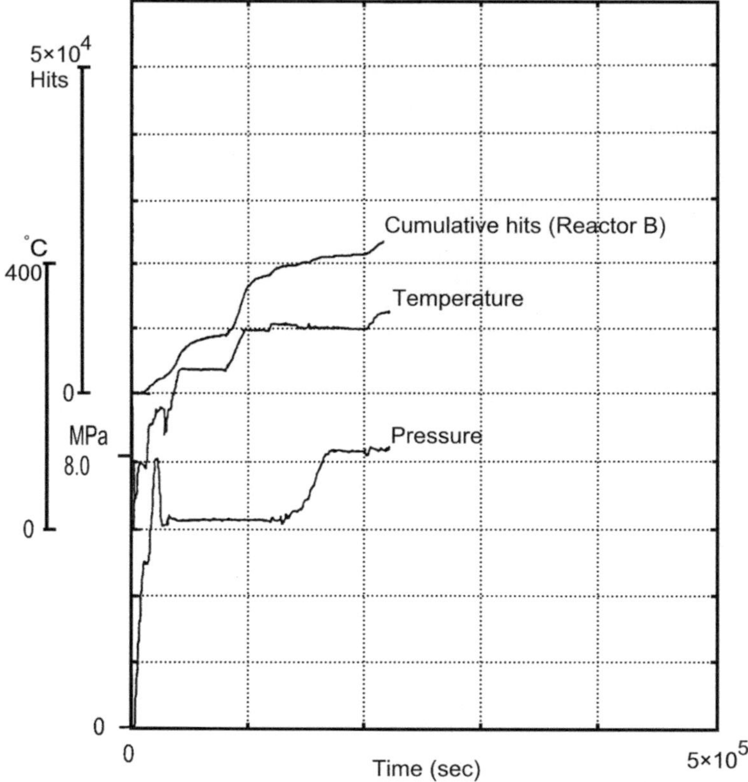

Fig. 6.3 Trends of cumulative AE hits, pressure, and temperature

As stated above, the application of AE testing to this type of actual structure is important because the integrity of the entire structure can be extensively verified by confirming the absence of significant AE.

6.2 Pipelines

Shigenori Yuyama

Long-distance pipelines exist in the states of Alaska, Utah, and Texas, USA, where AE testing has been widely applied as a practical inspection technique for effective maintenance. Advantages of the AE application include the ability to inspect a relatively long section of pipeline, giving location information on defects or leaks in one test, since a long sensor distance can be applied because of low attenuation in the pipeline.

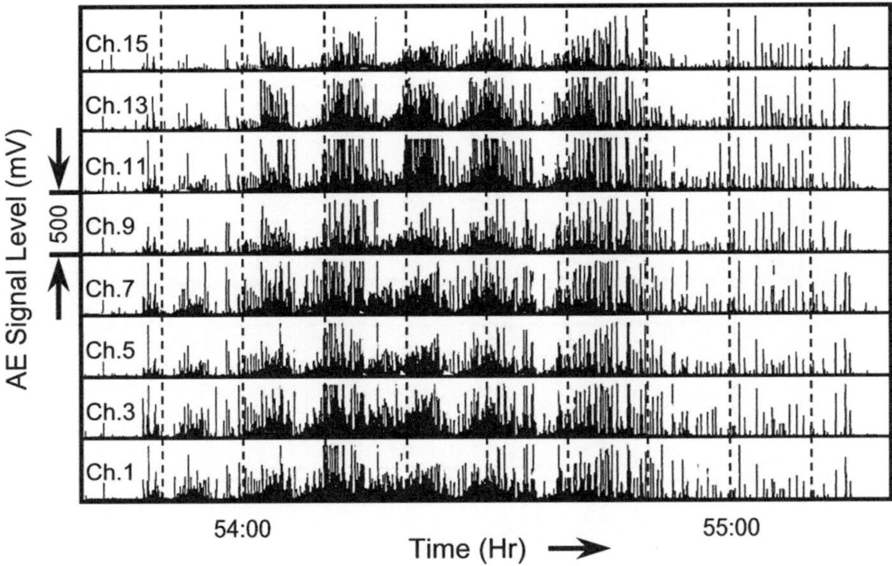

Fig. 6.4 Variation in the noise level

The sensor distance applicable to the testing depends on the type of product in the pipe (liquid or gas), the pressure inside the pipe, and the setup conditions of the pipe (aboveground or buried). If well-dried and high-pressure air is used for pressurization, the sensor distance can be extended to approximately 600 m.

AE testing is usually conducted for long-distance oil or natural gas pipelines. However, many cases have also been reported for short-distance pipelines in the transportation of naphtha or ammonia in chemical plants.

Recent works made in both the laboratory and field have reported useful information on AE behavior resulting from corrosion and fundamental tests of AE wave propagation in pipes. The applicability of AE testing to corrosion damage evaluation of buried pipes was investigated by comparing the results of AE testing with those from an ultrasonic test (UT) and visual test (VT). AE tests were conducted for 13 buried pipes in service in a refinery, as shown in Fig. 6.6, to make a comparison between the AE testing results and those of a UT and VT. Good correlation was found between the results of AE and other test methods. It has been reported that AE testing has been widely applied in refineries and chemical plants to evaluate corrosion damage in pipes.

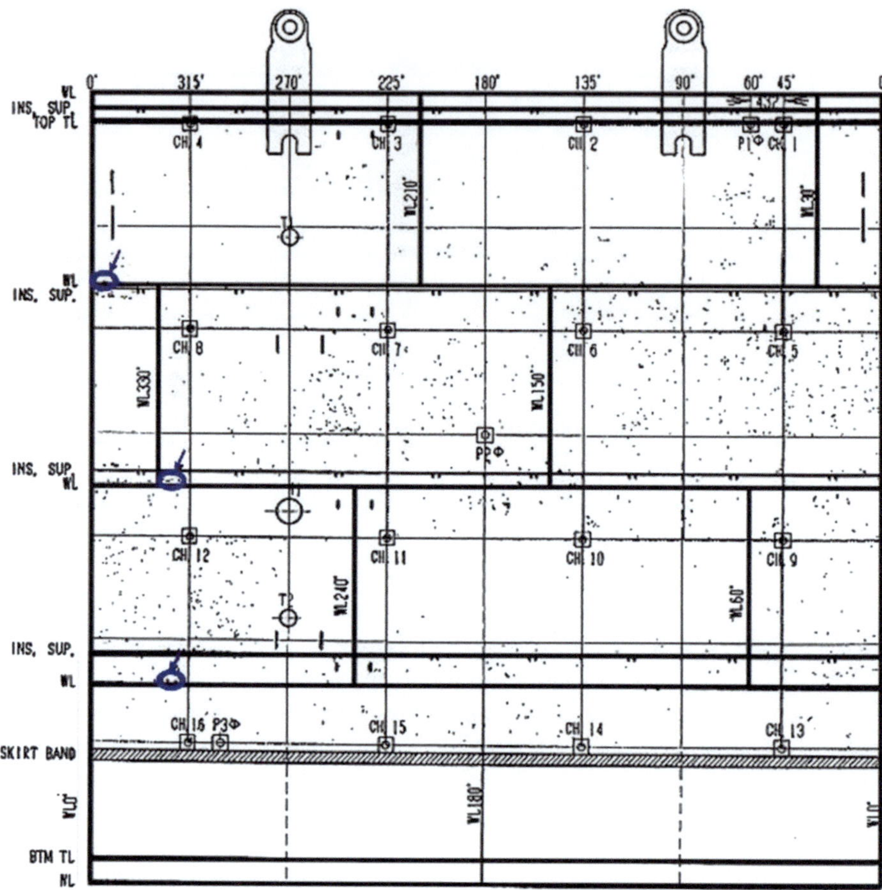

Fig. 6.5 Result of the AE source location (shell, Reactor B)

Table 6.2 Grade classification of AE sources

Grade	Activity	Number of AE events	Concentration of AE sources	Action
A	Very active	Many	High	Unload, then confirm by other NDTs
B	Active	Medium	Medium	Hold, or unload if necessary, then confirm by other NDTs
C	Slightly active	A few	Low	Continue AE test, record of results necessary
D	Not active	Very few	Sparse	Continue AE test, record of results not necessary

Fig. 6.6 Buried pipes in a refinery where the AE tests were conducted

6.3   High-Pressure Gas Storage Bottle

Shigenori Yuyama

In some states of the USA, AE testing is mandatory as an acceptance test, and periodic tests are performed every 5 years for high-pressure gas storage bottles of trailers. Linear (two-dimensional) source location with two AE sensors is applied and AE signals produced during pressurization to 110 % of the maximum operational pressure are detected and analyzed, as shown in Fig. 6.7. After the test, a UT is conducted for the locations where intense AE sources are detected by the source location. If defects with depth greater than 0.1 in. are found in the tested bottle, it will be discarded from operation according to the states' regulations.

Furthermore, the National Aeronautics and Space Administration (USA) has conducted AE tests of 120 or more gas storage vessels. The reason that AE testing has been applied to many containers is that it is a very time- and cost-effective test method for containers in service. Thus, the AE method has been widely employed as an inspection tool for pressure components. A pocket-type portable AE instrument widely used in the field is presented in Fig. 6.8.

6.4   Above Ground Storage Tank

Shigenori Yuyama

The bottom of above ground storage tank cannot be observed and inspected during operation. According to Japanese regulations, periodic internal inspection is required for tanks with storage capacity greater than 1000 kL. In the case of a

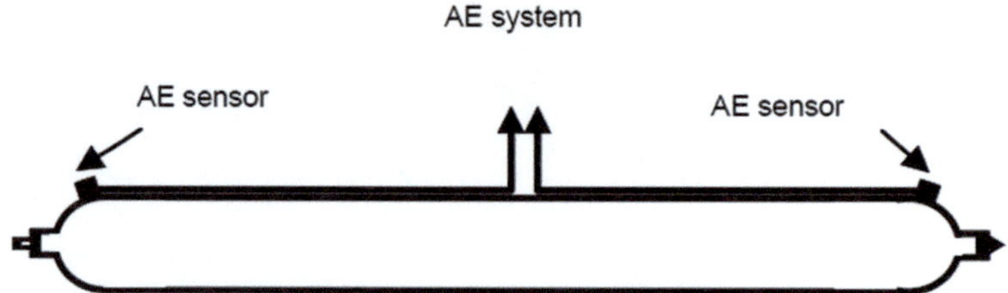

Fig. 6.7 Schematic representation of the AE testing of a high-pressure gas storage bottle

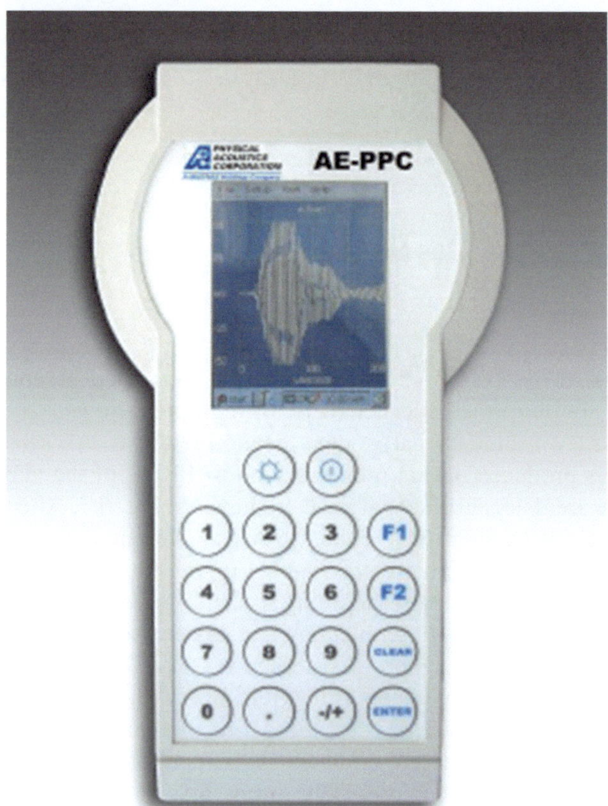

Fig. 6.8 Portable AE system

large above ground tank with capacity of 100,000 kL, it is very expensive to open the tank for internal inspection because of the high costs resulting from shut-down, cleaning, and inspection, which may often exceed several tens of millions of Yen.

An evaluation method has been developed in Europe and is based on a test procedure and a database consisting of data sheets for thousands of tests. In this method, multiple AE sensors are used to detect the AE signals resulting from active

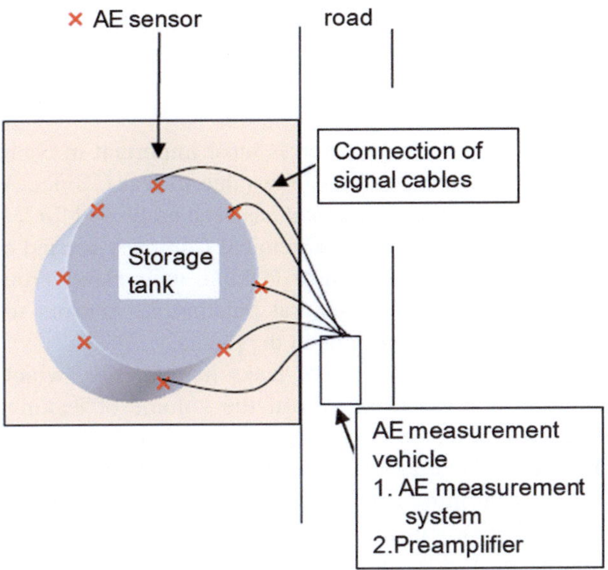

Fig. 6.9 Schematic illustration of the AE testing of a tank bottom

corrosion, making it possible to globally evaluate corrosion damage on a tank bottom.

As schematically shown in Fig. 6.9, AE sensors are placed at equal intervals in a circumferential direction on the tank wall at the height of 1–1.5 m above the tank floor. An experienced certified engineer collects AE data under the condition where no environmental noise due to rain and/or wind are detected, using an AE instrument installed in a vehicle parked outside an oil protection fence.

In European countries such as UK, France, Germany, the Netherlands, Italy, Spain and Greece, more than 1000 tanks are tested every year, while in the USA, a major oil company recently began an AE testing project as a part of AE application to risk-based inspection. It has been reported that about 500 tanks are tested annually in the USA. In Europe, this inspection method is in the process of being standardized by CEN (the European Committee for Standardization). Following these developments, AE testing for the evaluation of tank bottoms may be standardized in the very near future.

In Japan, the High Pressure Institute of Japan published a code (HPIS: Technical guideline for evaluation of corrosion damage to a tank bottom) with regard to AE testing of a tank bottom in May 2005. More than 250 tanks have been tested in accordance with the Japanese regulations and the AE results have been compared with thickness data obtained during internal inspection from either point measurements or floor scanning to confirm the applicability of the AE testing. Following the test procedure described in the HPIS code, several tens of tanks have been tested annually in Japan so far.

6.5   Leak Detection

Shigenori Yuyama

An effective technique for leak detection is most important in chemical plants to prevent serious accidents and financial losses due to leaks. Since AE sensors are very sensitive to leak noise, the AE method has been employed for leak detection in various fields. For instance, Monsanto Chemical Company carried out continuous AE monitoring for the early detection of leaks in their plants around the world, installing four or eight waterproof integral preamplifier sensors with a resonant frequency of 60 kHz at critical sections in the plants.

AE testing is used to evaluate valve leaks in refineries, which are a major concern. In the case of a gas leak, even if the volume of leaking gas is small, long-term leakage will result in enormous gas losses and huge economic losses. Therefore, it has been important to develop a reliable inspection technique that makes it possible to determine leaking valves among the hundreds of valves in a refinery and to quantitatively evaluate the volume of leaking gas.

At the beginning of 1980s, an oil company in UK developed an AE testing method that allows quantitative evaluation based on the correlation between detected AE data and leaks from valves. This method was developed in the following manner. First, a database was created for the relationship between the recorded AE data and the leaks, using artificially made leaks in different types of valves with different sizes under different pressures. This makes it possible to quantitatively evaluate the leaking volume by referring to the AE data. This AE technique has been used in several hundred refineries around the world in practical maintenance of valves as a daily inspection. Figure 6.10 demonstrates how the AE test is performed in a refinery.

6.6   Transformer

Shigenori Yuyama

A power station or substation has many large transformers, as shown in Fig. 6.11. Because of the aging of these units, an inspection technique is needed to evaluate the statuses of the units in a simple and reliable manner.

A partial discharge in a transformer has traditionally been evaluated by chemical analysis of sampled insulating oil collected from the transformer. However, the partial discharge generated in a transformer has long been known to produce detectable AE signals. Furthermore, it has been reported that oil gasification due to an increase in the local temperature of the insulating oil of a transformer also generates detectable AE signals. Thus, partial discharges and a local temperature increase of the transformer in service can be evaluated by monitoring AE signals.

Fig. 6.10 Detection and evaluation of a valve leak in a refinery using a portable leak monitor (*Photos courtesy of* www.mistrasgroup.com)

Fig. 6.11 Evaluation of partial discharge in a large transformer employing the AE method (*Photos courtesy of* www.mistrasgroup.com)

The partial discharge and temperature increase in a transformer should generate burst-type signals. AE source location using multiple sensors installed on the exterior wall of the transformer can be applied to evaluate approximate locations of the deteriorated areas. In addition, progress of the deterioration and operational conditions can be monitored using the AE data. The continuous AE monitoring of transformers with wireless AE sensors has been widely employed in power stations and sub-stations in the USA under a smart grid project.

6.7 Railway Bridges

Tomoki Shiotani

Aging of railway structures, most of which were constructed prior to other infrastructure, is becoming a serious problem, leading to maintenance problems in some cases. When an investigation of earthquake damage to a railway structure and a seismic diagnosis of such a structure are conducted, the presence, location, and degree of damage to the superstructure can be checked visually. On the other hand, the visual inspection of such underground substructures as foundations requires a higher cost and longer construction period for ground excavation and intervenes the running of in-service trains. For this reason, this type of visual inspection cannot actually be implemented. Accordingly, a useful method for investigating earthquake damage to bridge substructures based on AE activity was developed. The AE activity induced by the mobile load of an in-service train in the damaged area of the bridge structure is used to evaluate the damage of bridge.

As shown in Fig. 6.12, AE sensors are installed on the bridge pier to measure the AE activity generated from the defect in the bridge when the mobile load is imposed by the train. For this measurement, the AE source location is implemented to eliminate the train vibration noise and extract only useful AE signals. The Calm ratios, load ratios, amplitude distributions, and other parameters described in Chap. 5 for the AE events (Fig. 6.13) extracted by the source location are examined to estimate the level of deterioration of bridge.

For reference, Fig. 6.14 shows the workflow for evaluating the integrity of railway bridge piers by AE measurement. The necessity of continuous AE measuring is determined from the number of AE events. Then damage level of the bridge is estimated using b-values obtained from the amplitude distributions.

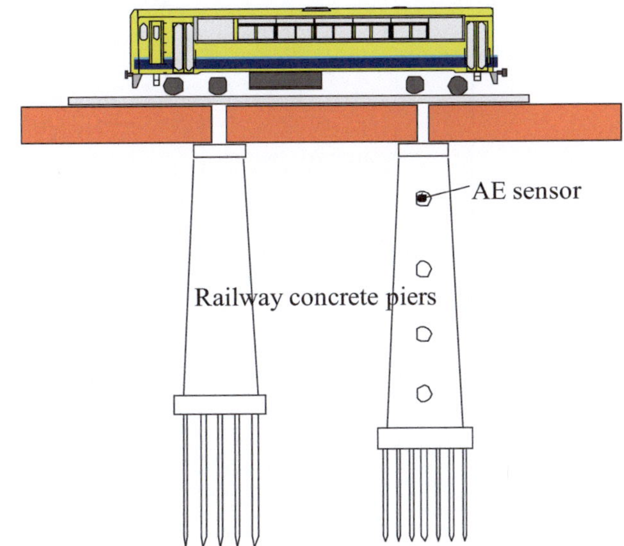

Fig. 6.12 AE monitoring of railway piers

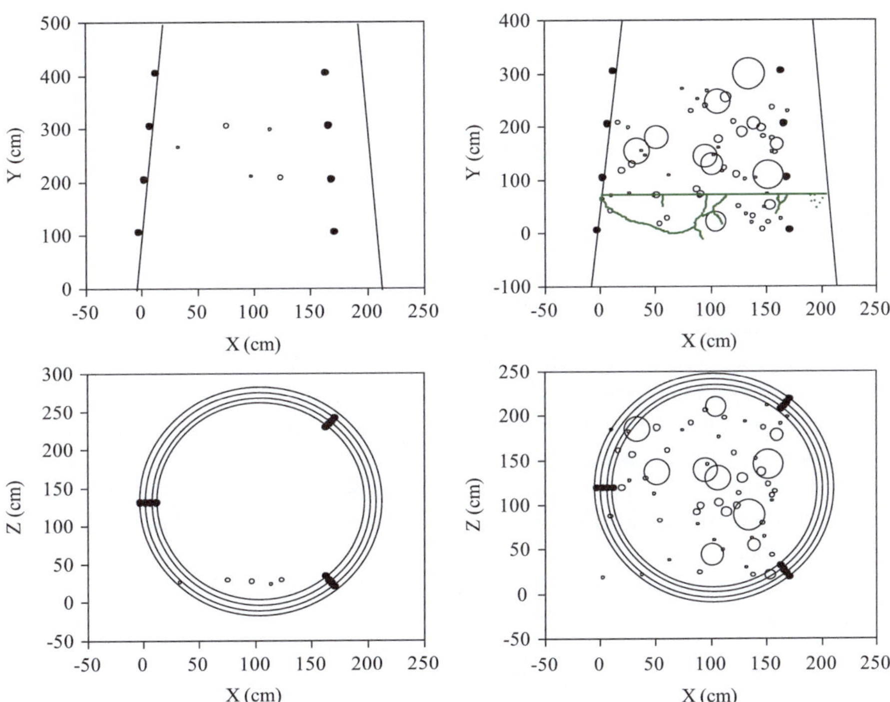

Fig. 6.13 AE events released from defects of railway piers (*left*: intact, *right*: seriously damaged)

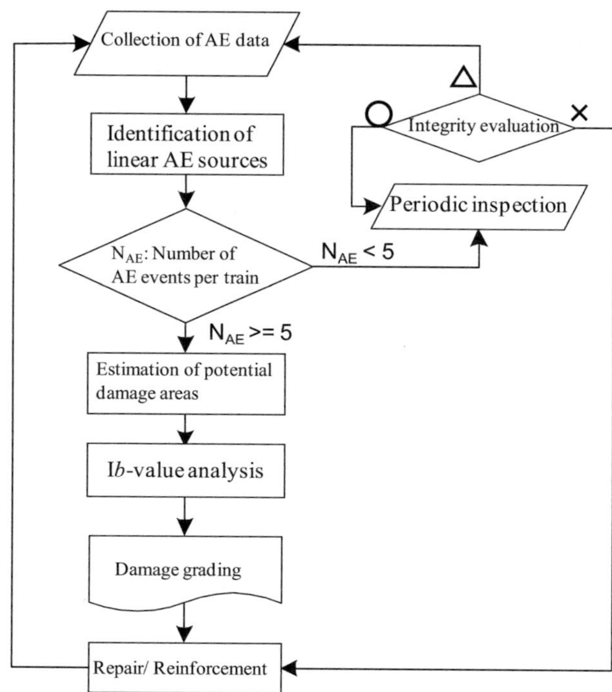

Fig. 6.14 Evaluation workflow with AE monitoring

6.8 Stability Monitoring of a Rock Slope

Tomoki Shiotani

Another field in which AE testing is expected to be employed to estimate the state of an inside fracture of materials and contribute to the prediction of final fracture is the stability monitoring of rock slopes. There are many challenges in measuring AE activity on rock slopes. A method to understand the AE activity generated in the inside of rocks and a method to eliminate the inevitable AE activity that is totally unrelated to rock deformations in the long-term measurements are needed. Therefore, a method was developed how to place AE sensors into a measurement borehole in the rock slope, accompanied with a reinforcing bar, filling the void in the borehole with cementitious materials that have the same physical properties as the actual rock.

In this method, the inside of the measurement borehole, as shown in Fig. 6.15, is replaced with cementitious materials. For this reason, existing cracks do not affect the propagation of AE waves. A local fracture along the existing cracks will generate an AE wave associated with the fracture of the filled materials. The detection of this AE wave by several AE sensors linearly arrayed on the reinforcing bar enables one-dimensional AE source location. Furthermore, it is possible to trace

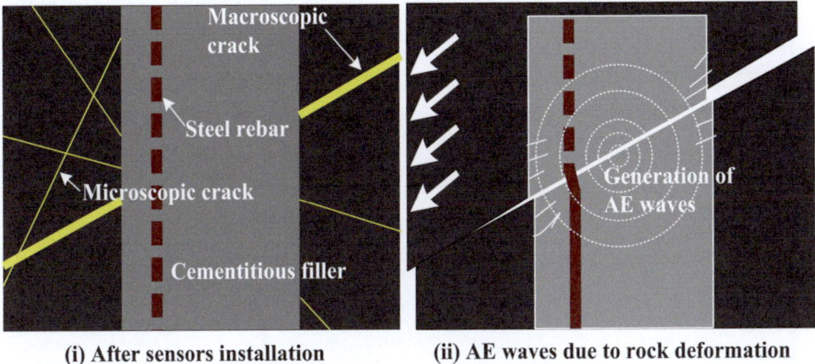

Fig. 6.15 Schematic illustration of the AE monitoring of rock

the fracturing process to eventual failure by tracking the AE activity generated from the friction between the reinforcing bar and filled materials.

Various types of fracture tests for different materials can be carried out in laboratory. Consequently, specific AE parameters can be determined according to the fracture patterns and levels. In this way, it becomes possible to reasonably evaluate the AE signals obtained from the actual rock slope on the basis of the fracture criterion obtained from the laboratory test. Figure 6.16 shows an example of AE application. AE sensors were installed on the reinforcing bar at intervals of 1.5 m. In this case, five AE sensors were placed within the AE measurement borehole, while cementitious materials with the same physical properties as the surrounding rock were used to fill the void in the rock.

6.9 Initial Ground Stress

Tomoki Shiotani

When artificial underground structures such as tunnels are constructed, it is important to estimate the inherent ground pressure of the rock, so as to monitor behavior of the surrounding rock and design necessary reinforcements for the structure. On the other hand, a phenomenon (the Kaiser effect) has been well known, showing that there is no AE activity until the load reaches the maximum load the material has previously experienced. Consequently, a method for estimating the initial ground stress based on the Kaiser effect has been established.

AE testing to estimate the initial ground stress is conducted through uniaxial compression tests of a fresh test specimen (tested within 3 days of being sampled) sampled in-situ (at the point to be measured). In this case, based on the number of AE hits or the cumulative number of AE hits generated at applied stresses, the stress that causes a remarkable increase in AE hits is taken as the estimated initial stress (see Fig. 6.17). It is important to note that in this AE-based test for the estimation of

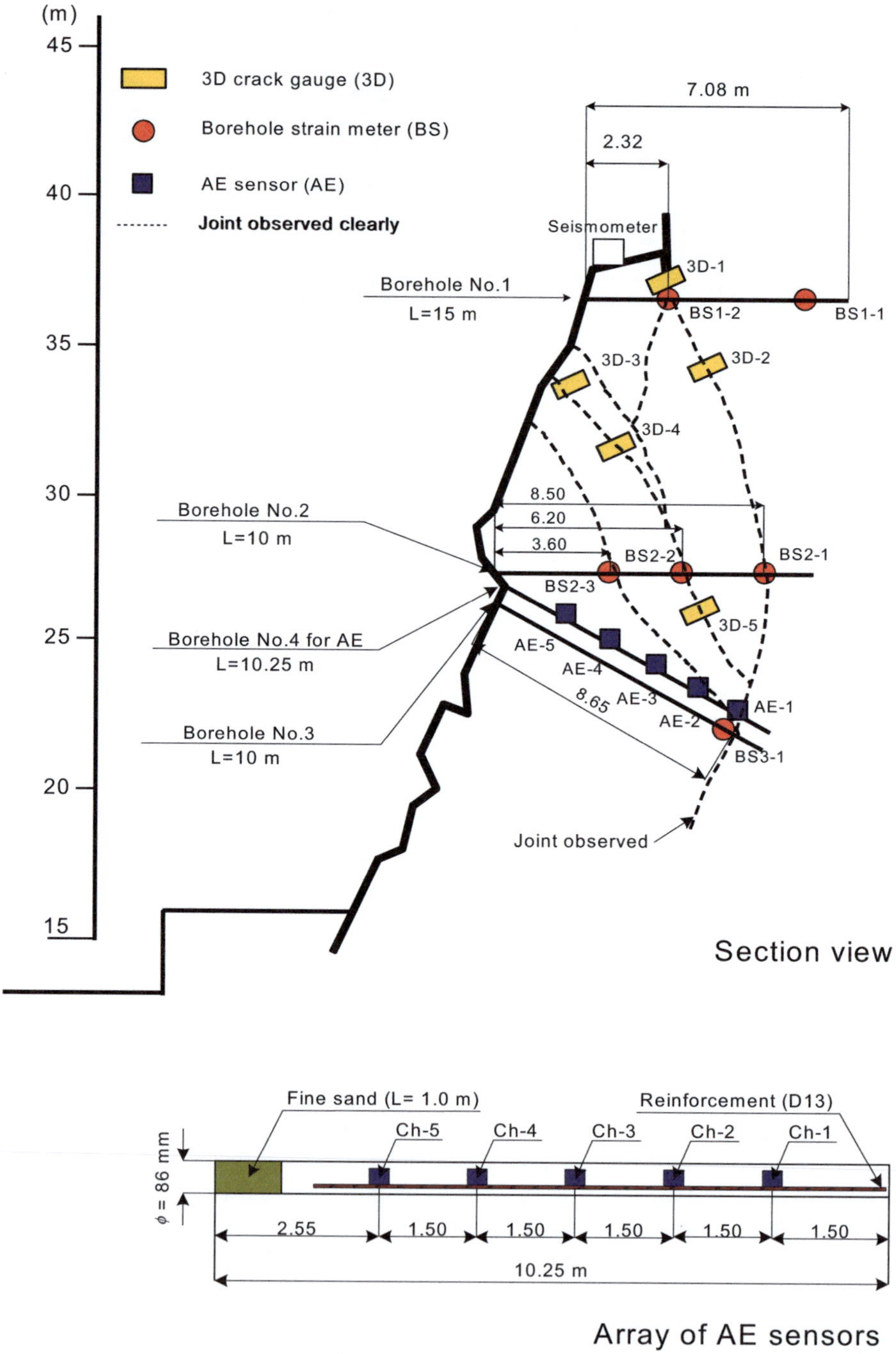

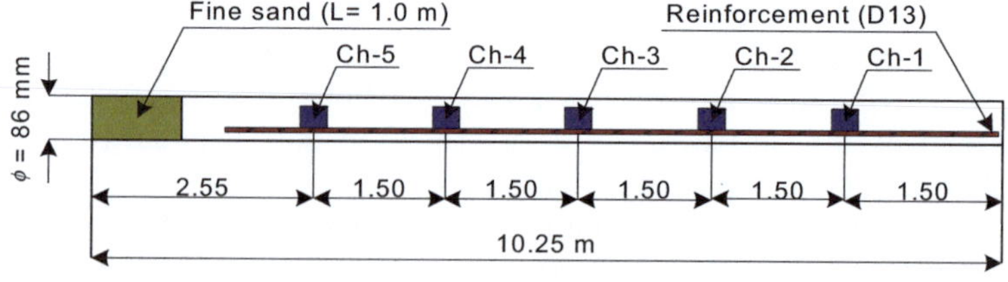

Fig. 6.16 Application of the AE monitoring of rock

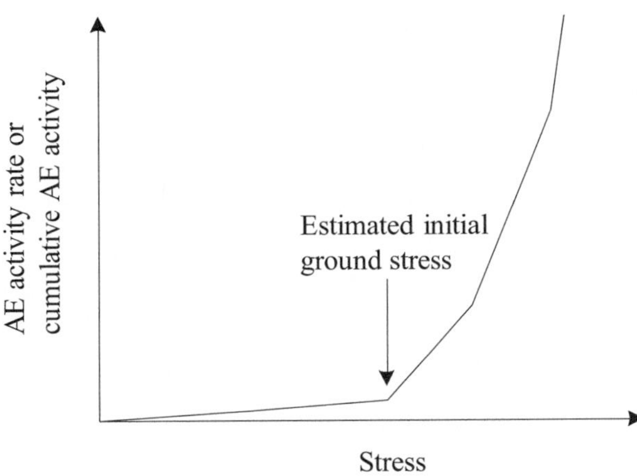

Fig. 6.17 Estimation of initial ground stress

initial ground stress, the estimated ground initial stress corresponds to the load-applied direction. Accordingly, when it is desirable to estimate the initial ground stresses in different directions, cores are sampled corresponding to these directions, and the loads are applied according to the sampling directions.

6.10 Aircraft

Sunao Sugimoto

The requirement of airframe structures is a high level of compatibility between weight saving and structural integrity for flight in air. Design concepts of the structures have changed repeatedly as a result of various accidents or problems. The present airframes are manufactured according to damage-tolerance design that ensures the remaining airframe life fully even if cracks are detected during an inspection. AE testing was used for the verification of airworthiness or integrity of the airframes as long as the design concept changes from traditional designs to the damage-tolerance design.

For example, General Dynamics F-111 fighter was manufactured before the introduction of the concept of damage-tolerant design. Although a safe-life structure in which there is no fatigue cracking in the airframes during the defined design life was adopted in this fighter design, the operation of the fighter had been suspended owing to a fatal failure accident before reaching the design service life. This has led to the implementation of the F-111 recovery program (G. Redmond, Proc. of 10th Asia-Pacific Conf. on NDT, 2001). A high load (-3 to $+7.33$ G) was applied to the airframe for 2–3 h at a temperature of $-43\,°C$. These loading tests were carried out to avoid decreasing the fracture toughness (to a more dangerous level) and the damage propagation to its secondary structure. The AE signals generated during the loading tests were used to evaluate the integrity of the

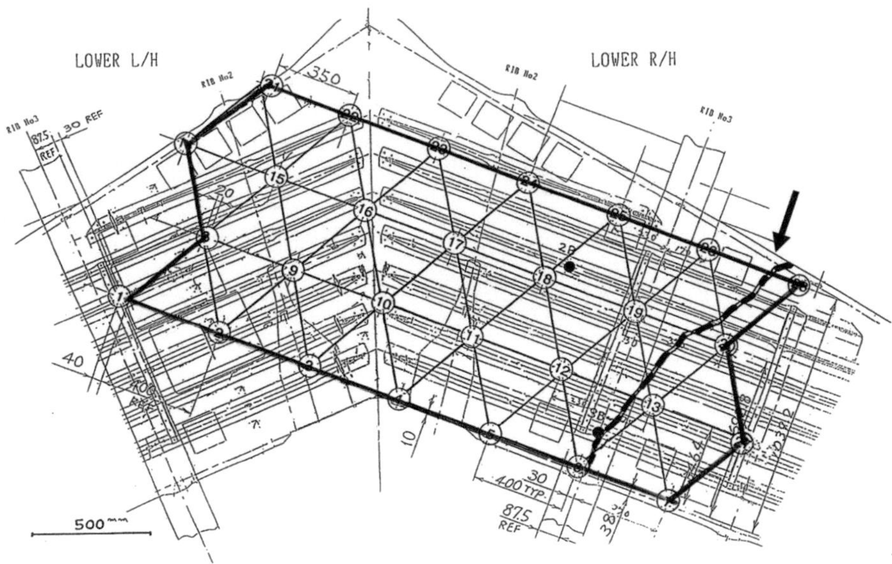

Fig. 6.18 AE sensor positions and crack path observed during structural test of a horizontal stabilizer in CFRP

fighter airframe. The test program has been implemented for new aircraft manufactured since 1969 to 1979, and shifted to Phase I (1973–1983), Phase II (1986–1998), and Phase III (1993–) as a structural inspection program for existing aircrafts.

Composite materials, particularly carbon fiber reinforced plastics (CFRPs), have increasingly been used for aircraft structures. Applications of the AE method to composite structures are also being promoted. A structural test was performed for a horizontal stabilizer in CFRP at the National Aerospace Laboratory of Japan (currently the Japan Aerospace Exploration Agency) as part of a joint development of a mid-sized civil aircraft by the Boeing Company of the USA and the Japan Aircraft Development Corporation (JADC), in which AE testing was employed for a residual strength test.

In this test, a monitoring area was divided into regular triangles with sides of 40 cm. Fig. 6.18 shows the overview of location of 28 AE sensors and crack path. An enlarged view of the source location result is shown in Fig. 6.19. When local damage occurred at 157 % of the limit load, the locations of the damage were confirmed by many AE sources in the area surrounded by three AE sensors: No. 6, No. 12 and No. 13 sensors. Furthermore, numerous AE signals were detected in the area surrounded by four AE sensors: No. 19, No. 20, No. 26 and No. 28 sensors at 167 % of the limit load. The interesting result is that the final failure occurred along the line connecting these two damage areas.

Besides the above mentioned examples, AE testing has been applied to a wide variety of aircrafts and military aircrafts because they are operated under severe

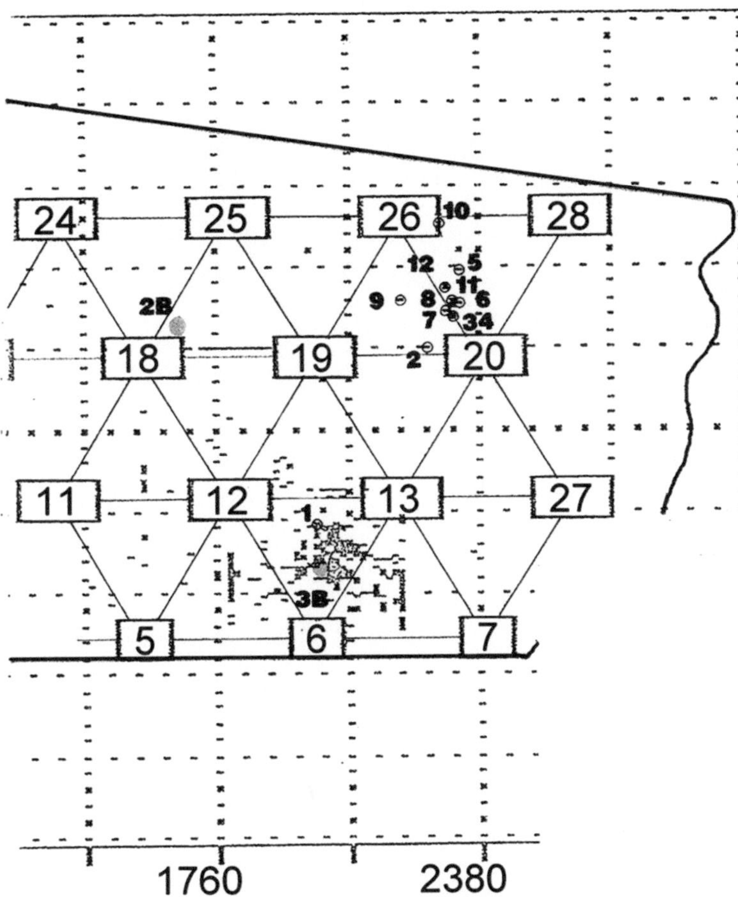

Fig. 6.19 AE source location observed during structural test of a horizontal stabilizer in CFRP

temperature and load conditions. For example, although landing gears are not damage tolerance structures generally, their operation has been optimized using an AE measurement system. There has been an increase in such research into the structural health monitoring of aircraft, and the AE technique is expected to contribute to further developments in this field.

Glossary

Precautions

1. This Glossary is presented to help beginners learn about acoustic emission testing. The definitions for the terms adhere to JIS (Japanese Industrial Standards) and NDIS (JSNDI Standards), but additions to the wording and partial omissions were made to facilitate understanding.
2. The symbol "→" refers to an opposite or related term. After searching for a target term, any associated term marked with → can be looked up to gain a better understanding.
3. Different technical terms are used in related industries. Although this is true in relation to the terms used for this glossary adopts the terms most widely used as idiomatic expressions in the industry. Consequently, some of the terms differ from those specified in JIS and NDIS.

Acoustic emission (AE) Phenomenon that produces elastic waves through the release of the strain energy accumulated when a solid is deformed or fractured, or the transient elastic wave generated in this way

Acoustic emission event →event

Acoustic emission testing (AET, AE testing, AT) Non-destructive testing and material evaluation test conducted with the use of AE

Acousto-ultrasonic (AU) Non-destructive testing method used to generate elastic waves to detect and evaluate the distribution of flaws in a structure, severity of the damage, and any change in the mechanical characteristics of the test piece

Note: This AE inspection method is a combination of an AE signal-based analysis method and an ultrasonic material properties testing method.

AE →Acoustic emission

© Springer Japan 2016
The Japanese Society for Non-Destructive Inspection, *Practical Acoustic Emission Testing*, DOI 10.1007/978-4-431-55072-3

AE channel A measuring system with a combination of the following devices and cable:

(1) AE sensor
(2) Preamplifier or impedance matching transducer
(3) Filter, secondary amplifier, or other necessary equipment
(4) Connection cable
(5) Detector or processor

AE count (ring down count, count, emission count) The number of times that an AE signal exceeds a preset threshold during any selected portion of a test

AE count rate (emission rate, count rate) AE counts per unit time

AE energy A value determined after an instantaneous value squared is integrated; the maximum amplitude squared, or the maximum amplitude multiplied by a duration is used as a simple value; note that the definition depends on the device

AE event →Event

AE root-mean-square (RMS) value Effective value of an AE signal

AE sensor (AE transducer) Converter that converts AE waves to electric signals.

AE signal Electric signal that is converted from an AE wave with an AE sensor

AE signal amplitude →Amplitude

AE signal duration →duration

AE signal end Recognized termination of an AE signal, usually defined as the last crossing of the threshold by that signal

AE signal maximum amplitude →Amplitude

AE signal peak amplitude →Amplitude

AE signal rise time →Rise time

AE signal start Beginning of an AE signal as recognized by the system processor, usually defined by an amplitude excursion exceeding threshold

AE source Source where AE is generated

AE testing →Acoustic emission testing

AE wave Elastic wave generated by AE

AE waveguide (waveguide) →waveguide

Amplitude (AE signal amplitude, AE signal maximum amplitude, AE signal peak amplitude) Maximum voltage of an AE signal waveform in one emission event; it sometimes refers to the maximum of an absolute value in an envelope for AE signals

Arrival time difference The difference in the arrival times that AE waves reach several AE sensors

Artificial AE source Source of elastic waves simulating AE waves that are used for the calibration and sensitivity setting of an AE sensor or measuring instrument

ASL →Average signal level

Attenuation Attenuation of amplitude due to the absorption and diffusion of AE waves when the waves propagate through a medium

Note: This is normally expressed as a dB value per unit length

AU →Acousto-ultrasonic method

Average signal level (ASL) Rectified and time-averaged logarithmic AE signal. Value measured on a logarithmic scale for AE amplitudes and reported as a unit of dB_{AE}; 1 μV (micro volt) is defined as 0 dB_{AE} at the input terminal of a preamplifier

Burst AE (Burst emission, Transient AE, Transient emission) AE signal that can be apparently divided on a temporal basis

Burst emission →Burst AE

Continuous AE (Continuous emission) AE signal that apparently cannot be divided on a temporal basis

Continuous emission →Continuous AE

Couplant Material used at the interface between a structure and an AE sensor for the smooth transmission of acoustic energy in AE monitoring

Cumulative AE amplitude distribution →Cumulative amplitude distribution

Cumulative amplitude distribution (Cumulative AE amplitude distribution) Number of AE events with signals that exceed arbitrary amplitudes as a function of amplitude V

db_{AE} Logarithmic value of the AE signal amplitude relative to 1 μV (micro volt); it is expressed as the peak amplitude of the signal $[(dB_{AE}) = 20 \log 10 \, (A_1/A_0)]$, where

A_0: is equal to 1 μV at the sensor output (before amplification) and

A_1: is the peak voltage of the measured AE signal.

Dead time Any interval during data acquisition when a measuring instrument or system cannot receive new data

Differential amplitude distribution (Differential AE amplitude distribution) Number of AE events with signal amplitudes between amplitudes of V and V $+ \Delta V$ as a function of the amplitude V, where f(V) is the absolute value of the derivative of the cumulative amplitude distribution F(V)

Duration (AE signal duration) Time from the start to end of an AE signal

Dynamic range The difference between the overload level and minimum signal level in a system or sensor as expressed in dB

Emission event →Event

Evaluation threshold Threshold used for the analysis of data after testing; the threshold is set to the same value as the voltage threshold in most cases

Event (AE event, acoustic emission event, emission event) Local material change giving rise to acoustic emission

Event count (AE event count) The number obtained by counting each discernable acoustic emission event once

Event count rate (AE event count rate) Event count per unit time

Hit (AE hit) Any signal that exceeds a threshold and provides system channel data

Kaiser effect Absence of detectable AE at a fixed trigger level until previously applied stress levels are exceeded

Linear source location →One-dimensional source location

One-dimensional source location (linear source location) The determination of a one-dimensional source location requiring two channels or more

Planar source location →Two-dimensional source location

Ring down count →AE count

Rise time (AE signal rise time) Time interval between an AE signal start and the peak amplitude of that AE signal

RMS →AE Root mean square value

SN ratio Ratio of the signal amplitude to the noise level

Source location (AE source location) →One-dimensional location, two-dimensional location, three-dimensional location

Method for determining the position of AE sources in a structure

Three-dimensional source location The determination of a three-dimensional source location requiring four channels or more; 5 channels or more are generally used

Transient AE →Burst AE

Transient emission →Burst AE

Two-dimensional source location (planar source location) The determination of a two-dimensional source location requiring three channels or more

Voltage threshold Voltage above which a signal is recognized; this voltage threshold can be adjusted and fixed by users or made an automatic floating type

Waveguide (AE waveguide) Device that couples elastic energy from a structure or other test object to a remotely mounted sensor during AE monitoring

Note: An example of an AE waveguide would be a solid wire or rod that is coupled to a monitored structure at one end and to a sensor at the other end.

Zone location Method to determine the general region of an AE source (using, for example, the total AE counts, energy, or hits)

Note: Several approaches to zone location are employed, including independent channel zone location, first-hit zone location and arrival sequence zone location.

Appendix: List of Codes and Standards

Introduction

Acoustic emission testing is widely used for materials testing, the evaluation of structural integrity and the continuous monitoring of structures. Numerous codes and standards have been issued so far in Japan and other countries, providing guidelines for sensor calibration, instrument evaluation, test procedures, and data collection.

Codes and Standards

1. ISO 12713: "Non-destructive testing – Acoustic emission inspection – Primary calibration of transducers"
2. ISO 12714: "Non-destructive testing – Acoustic emission inspection – Secondary calibration of acoustic emission sensors"
3. ISO 12716: "Non-destructive testing – Acoustic emission inspection – Vocabulary"
4. ISO 9712: "Non-destructive testing – Qualification and certification of personnel"
5. ISO/DIS 16148: "Gas cylinders – Refillable seamless steel gas cylinders and tubes – Acoustic emission examination (AT) and follow-up ultrasonic examination (UT) for periodic inspection and testing"
6. ISO TR 13115: "Non-destructive testing-Methods for absolute calibration of acoustic emission transducers by reciprocity technique"
7. ISO TR 25107: "Non-destructive testing – Guidelines for NDT training syllabuses" (under discussion)
8. ISO TR 25108: "Non-destructive testing – Guidelines for NDT personnel training organizations" (under discussion)

© Springer Japan 2016
The Japanese Society for Non-Destructive Inspection, *Practical Acoustic Emission Testing*, DOI 10.1007/978-4-431-55072-3

9. ASME Boiler and Pressure Vessel Code: "Acoustic Emission Examination of Fiber-Reinforced Plastic Vessels" Section V, Article 11
10. ASME Boiler and Pressure Vessel Code: "Acoustic emission examination of metallic vessels during pressure testing" Section V, Article 12
11. ASME Boiler and Pressure Vessel Code: "Continuous Acoustic Emission Monitoring" Section V, Article 13
12. Recommended practice No. SNT-TC-1A*, Personnel qualification and certification in nondestructive testing, ASNT, 1996
13. ASTM E 569-76: "Standard recommended practice for acoustic emission monitoring of structures during controlled stimulation"
14. ASTM E1888/E1888M-02: "Acoustic Emission Examination of Pressurized Containers Made of Fiberglass Reinforced Plastic with Balsa Wood Cores"
15. ASTM E2076-05: "Examination of Fiberglass Reinforced Plastic Fan Blades Using Acoustic Emission"
16. ASTM E2191-02: "Examination of Gas-Filled Filament-Wound Composite Pressure Vessels Using Acoustic Emission"
17. ASTM E1930-02: "Examination of Liquid Filled Atmospheric and Low Pressure Metal Storage Tanks Using Acoustic Emission"
18. ASTM E1419-02b: "Examination of Seamless, Gas-Filled, Pressure Vessels Using Acoustic Emission"
19. ASTM E1106-86(2002): "Primary Calibration of Acoustic Emission Sensors"
20. ASTM E1067-01: "Acoustic Emission Examination of Fiberglass Reinforced Plastic Resin (FRP) Tanks/Vessels"
21. ASTM E1118-05: "Acoustic Emission Examination of Reinforced Thermosetting Resin Pipe (RTRP)"
22. ASTM E749-01: "Acoustic Emission Monitoring During Continuous Welding"
23. ASTM E751-01: "Acoustic Emission Monitoring During Resistance Spot-Welding"
24. ASTM E569-02: "Acoustic Emission Monitoring of Structures During Controlled Stimulation"
25. ASTM E1736-05: "Acousto-Ultrasonic Assessment of Filament-Wound Pressure Vessels"
26. ASTM E750-04: "Characterizing Acoustic Emission Instrumentation"
27. ASTM E1139-02: "Continuous Monitoring of Acoustic Emission from Metal Pressure Boundaries"
28. ASTM E1211-02: "Leak Detection and Location Using Surface-Mounted Acoustic Emission Sensors"
29. ASTM E1781-98(2003): "Secondary Calibration of Acoustic Emission Sensors"
30. ASTM E2075-05: "Verifying the Consistency of AE-Sensor Response Using an Acrylic Rod"
31. ASTM E1932-97(200): "Acoustic Emission Examination of Small Parts"
32. ASTM E2374-04: "Acoustic Emission System Performance Verification"

33. ASTM E1495-02: "Acousto-Ultrasonic Assessment of Composites, Laminates, and Bonded Joints"
34. ASTM E976-00: "Determining the Reproducibility of Acoustic Emission Sensor Response"
35. ASTM E650-97(2002): "Mounting Piezoelectric Acoustic Emission Sensors"
36. DIN EN 14584: "Non-Destructive Testing – Acoustic Emission – Examination of Metallic Pressure Equipment during Proof Testing; Planar Location of AE Sources"
37. EN 1330-9: "Non-Destructive Testing – Terminology – Part 9, Terms Used in Acoustic Emission Testing"
38. EN 13477-1: "Non-Destructive Testing – Acoustic Emission – Equipment Characterization – Part 1, Equipment Description"
39. EN 13477-2: "Non-Destructive Testing – Acoustic Emission – Equipment Characterization – Part 2, Verification of Operating Characteristics"
40. EN 13554: "Non-Destructive Testing – Acoustic Emission – General Principles"
41. JIS Z 2342: Method for acoustic emission testing of pressure vessels during pressure tests and classification of test results
42. NIDS 2106: Methods for assessing the performance characteristics of an acoustic emission testing system
43. NDIS 2109: Method for absolute calibration of acoustic emission transducers employing reciprocity
44. NDIS 2110: Method for measuring the sensitivity degradation of an acoustic emission transducer
45. NDIS 2419: Recommended practice for continuous monitoring of metal pressure vessel by acoustic emission
46. NDIS2421: Recommended practice for in situ monitoring of concrete structures by acoustic emission
47. HPIS G 110 TR 2005: Recommended practice for acoustic emission evaluation of corrosion damage to the8bottom plate of oil storage tanks
48. HPIS E 102 TR 2012: Recommended practice for acoustic emission evaluation of corrosion damages in underground tanks

Index

© Springer Japan 2016

The Japanese Society for Non-Destructive Inspection, *Practical Acoustic Emission
Testing*, DOI 10.1007/978-4-431-55072-3

MIX
Papier aus verantwortungsvollen Quellen
Paper from responsible sources
FSC® C105338

If you have any concerns about our products,
you can contact us on
ProductSafety@springernature.com

In case Publisher is established outside the EU,
the EU authorized representative is:
Springer Nature Customer Service Center GmbH
Europaplatz 3, 69115 Heidelberg, Germany

Printed by Libri Plureos GmbH
in Hamburg, Germany